WORLD BANK TECHNICAL PAPER NO. 365

Investing in Pastoralism

Sustainable Natural Resource Use in Arid Africa and the Middle East

David John Pratt, François Le Gall, and Cornelis de Haan

The World Bank
Washington, D.C.

First printing June 1997

Technical Papers are published to communicate the results of the Bank's work to the development community with the least possible delay. The typescript of this paper therefore has not been prepared in accordance with the procedures appropriate to formal printed texts, and the World Bank accepts no responsibility for errors. Some sources cited in this paper may be informal documents that are not readily available.

The findings, interpretations, and conclusions expressed in this paper are entirely those of the author(s) and should not be attributed in any manner to the World Bank, to its affiliated organizations, or to members of its Board of Executive Directors or the countries they represent. The World Bank does not guarantee the accuracy of the data included in this publication and accepts no responsibility whatsoever for any consequence of their use. The boundaries, colors, denominations, and other information shown on any map in this volume do not imply on the part of the World Bank Group any judgment on the legal status of any territory or the endorsement or acceptance of such boundaries.

ISBN 0-8213-3943-5
ISSN: 0253-7494

David John Pratt is a consultant at the World Bank. François Le Gall is a consultant/livestock specialist in the Agriculture and Natural Resources Department of the World Bank. Cornelis de Haan is a livestock adviser in the Agriculture and Natural Resources Department of the World Bank.

Library of Congress Cataloging-in-Publication Data

Pratt, David, 1932–
Investing in pastoralism : sustainable natural resource use in arid Africa and the Middle East / David Pratt, François Le Gall, and Cornelis de Haan.
p. cm. — (World Bank technical paper ; no. 365)
Includes bibliographical references (p).
ISBN 0-8213-3943-5
1. Pastures—Developing countries—Planning. 2. Animal industry—Developing countries—Planning. 3. Agricultural development projects—Developing countries—Planning. 4. Arid regions agriculture—Developing countries—Planning. 5. Natural resources, Communal—Developing countries—Planning. 6. Conservation of natural resources—Developing countries—Planning. I. Le Gall, François, 1961– . II. Haan, C. de. III. World Bank. IV. Title.
HD1641.D44P7 1997
333.74'0915'4—dc21 97-21677
CIP

Contents

Foreword

About one-third of the world's productive surface is covered by arid or semi-arid ecosystems, where pastoral production is the main — and often the only — activity. Pastoral production, however, is coming under increasing stress. Human population growth often exceeds the rate of increase in animal production in these areas, and population pressure fuels the encroachment of crop farming onto the most valuable grazing areas. Social deprivation is therefore acute, and while the arid ecosystems have shown high resilience, the sustainability of rangelands remains a key issue. Pastoral production can therefore not be ignored on the development agenda.

The World Bank has been involved in range livestock development since the 1970s, and much has been learned in the process. In addition, a new body of knowledge has become available over the last decade, which also contributes to what are currently considered good practices in natural resource management in pastoral areas. This document provides an overview of these good practices, covering the enabling environment, requirements for overall project preparation, and individual project components.

The Agriculture and Natural Resources Department of the World Bank has undertaken a general assessment of available technologies and institutional arrangements over the last few years. This paper is part of that assessment of natural resource management in arid and semi-arid ecosystems, and offers suggestions about integrating this knowledge into World Bank lending operations. It is directed at all those involved in the promotion of sustainable development. We are hopeful that the information in this document will be useful to those in a position to improve the conservation of the resource base in one of the most difficult and complex environments on earth, and improve the livelihood of those who live in these regions.

Alex F. McCalla
Director
Agriculture and Natural Resource Department

Acknowledgments

The authors are grateful for the inputs and contributions from many individuals in the preparation of this document. Douglas Forno originally recognized the need to update our understanding of good practices in pastoral development, and the Sahelian Operational Review (SOR) provided valuable support for field review of pastoral activities in African arid and semi-arid rangelands. Jeremy Swift from the Institute of Development Studies, Sussex, UK, and Walter Lusigi and Laurent Msellati of the World Bank were the peer reviewers. Seth Beckerman provided all the editorial support.

Preface

This document focuses on natural resource management (NRM) on arid rangelands used by pastoralists in Africa and the Middle East. Pastoralism in other regions — particularly southwest and central Asia — is also discussed, but without the detailed analysis of these other systems. Similarly, although agropastoral situations and grazing orbits that extend into wetter zones are covered, the principal focus is areas where there are only 30 to 90 growing days per year. This usually equates to an annual rainfall of 150 to 450 millimeters. Rainfed cropping is extremely risky in these areas, and inhabitants draw all or most of their subsistence from livestock.

Although pastoralists keep livestock, pastoralism is more than a system of livestock production. Livestock are the currency of social cohesion as well as of subsistence and trade. Although the influence of pastoralists is much reduced from the days when pastoralism was a dominant geopolitical system over much of central Asia, eastern Europe, the Middle East, and Africa, pastoralism is still widely practiced on arid lands. Wherever it occurs, sustainable use of rangeland resources remains a key issue.

At least 10 million square kilometers can be categorized as arid or semi-arid rangeland. This area doubles when sparsely used deserts and semi-deserts are included. At present, these rangelands provide a direct livelihood for about 180 million people living in close association with about 360 million cattle and over 600 million sheep and goats (Sere, Steinfeld, and Groenewold 1996). Moreover, social deprivation is acute, and these areas are perceived as being among the most degraded of any in the world (Dregne, Kassas, and Rozanov 1992).

Although the World Bank started funding NRM projects only recently, it has been supporting range livestock development since the 1960s. Although these early projects contributed little to pastoral development, much has been learned in the process. Present-day projects can also draw from a new body of literature on range ecology and pastoral development (Behnke, Scoones, and Kerven 1993; Scoones 1994), including case studies such as that by Lane (1996). Early lessons, new literature, and recent experience in the application of holistic resource management (Savory 1988) have all contributed to the current appreciation of what constitute best practices in NRM in pastoral areas.

The core of pastoral development is support for sustainable NRM coupled with improved social services, particularly education and health. NRM in pastoral areas, however, is more complex than is sometimes appreciated. Although easily comprehended as a general concept or objective, NRM is always the result of interaction among three quite distinct systems — resources, resource users, and the larger geopolitical system in which they operate — each of which needs to be understood if intervention is to be effective.

Executive Summary

Abstract

This document offers guidelines for development in arid lands where pastoralism is practiced. The focus is on natural resource management (NRM), which is a prime consideration in pastoral development. An introductory chapter on the nature of NRM is followed by advice on preparing for project intervention (Part One), and guidelines for specific project components (Part Two). A concluding chapter considers the broader implications for international agencies such as the World Bank. Eight annexes provide additional background information and advice, and a user guide is offered for the practioner.

Key words: Pastoralism – Arid lands – Rangeland – Livestock – Africa – Middle East – Natural resources – Mobility (in resource use) – Population pressure – Development planning – Institutions – Drought management – Technology – Project design – Monitoring

Summary

Introduction (Chapter 1). NRM is more complex than is commonly appreciated. It is the product of interaction among the natural system that produces the resources on offer, the user system that is exploiting these resources, and the wider geopolitical system in which the resource users are operating. Effective intervention in NRM therefore requires equal familiarity with the workings of each of these systems and how they interact. The task is further complicated by the unusual nature of pastoralism and the attitudinal problems that surround it. These aspects are all summarized in this chapter with further background information on customary NRM, past development efforts, and new concepts in NRM (provided in Annexes A, B, and C).

Part One. Preparing for Project Intervention. Effective pastoral development has four basic requirements:

Differentiate the type of pastoral system(s) being assisted by reference to mobility, livestock species, economic orientation (milk, meat, trade), involvement in other enterprises, and relationship to external economies and socio-territorial organization (Chapter 2 and Annex D). The need to categorize pastoral systems is emphasized because so much of the writing on pastoralism is too general to be useful for effective development. Yet the attributes mentioned above — not least the value systems and the economic orientation of any pastoral system — radically affect the choice of development inputs.

Assess population pressure and societal coping mechanisms in order to establish whether present population pressure represents a serious impediment to progress (Chapter 3). Indicators of pressure fall into three categories — societal indicators assess the effectiveness of welfare mechanisms, household indicators lie principally in stock wealth, and environmental indicators lie in the total landscape and its use.

Ensure that enabling policies and infrastructure are in place to support pastoral development (Chapter 4 on Policy Issues and Chapter 5 on the Institutional Framework). The starting point is to recognize the intrinsic value of pastoral economies and values, and that they are affected, just like any other sector, by macroeconomic conditions (exchange rate, trade and incentive policies). The policy decision to invest in pastoralism is then based on economic, social, environmental, and policy considerations

Ensure a project design that combines participation, flexibility, and the prospect of sustainability (Chapter 6 and Annex E on Participatory Rural Appraisal Methodologies). It is important that the project embraces the entire area used by the pastoral population (including dry-season grazing areas), involves the target population in the preparatory and the implementation phase, and bases project design on an overall assessment of the development goals and the most likely development path to achieve those goals.

Part Two. Guidelines for Specific Project Components. The guidelines address five essentials of pastoral development projects.

Herder Organizations (Chapter 7) have a central role in all pastoral development. However, it is important that forms of organization — whether customary or introduced — fit the tasks expected of them. These tasks can vary from NRM (principally grazing and water management), veterinary and medical supplies, borehole operation, marketing, and advocacy functions. Organizational requirements for these different functions vary, and emphasis in the guidelines is therefore given to differentiating likely functions and guiding the structure of local organizations to suit those functions.

Support Systems (Chapter 8) are required to enable herder organizations (and the whole development process) to operate effectively. Central government, local government, and the commercial private sector all have a role, and the guidelines focus on how best to allocate and share that responsibility. The present trend is toward cost recovery and devolving service delivery toward the private sector, with central government focusing on policy and control and providing essential infrastructure.

Drought Management (Chapter 9) must feature as a principal project component or as a feature permeating all components. The starting point is to recognize drought as a condition of *unexpectedly* low rainfall (since misdiagnosis of feed shortages and human suffering does not help effective treatment), and to appreciate indigenous drought strategies in relation to different drought intensities. The help that is given will vary among projects, but there are also commonalties in packaging assistance that are elaborated in the text.

Phasing of Technical Inputs (Chapter 10) is as important as the choice of the technologies to introduce. The relevant technologies — water development, range management, animal health and production, commerce and marketing systems, wild life utilization, and crop land improvement — must be carefully sequenced. This sequencing needs to relate to the development path that is projected, as well as to environmental sensitivities and management capabilities. New water supplies, for example, can be as environmentally disruptive as they are helpful. Three phases of inputs are elaborated.

Process Monitoring (Chapter 11) must be incorporated as a distinct project component in order to ensure the flow of information needed to guide the evolution of pastoral development. Such monitoring is the characteristic of 'process projects' which are seen as essential for pastoral development (Chapter 6). Guidelines are included for monitoring in each of the parent systems (natural, user, and geopolitical) that govern NRM, and additional guidance is included in Annex H.

Conclusion (Chapter 12). Application of the guidelines presented above leads development agencies into procedures and activities outside those normal for sectoral projects. This final chapter examines four of the broader aspects of aiding pastoral development:

- management issues associated with the process approach;
- improving baseline knowledge of pastoral systems and their environments;
- assisting technology development; and
- overcoming attitudinal problems about pastoralism.

This field is potentially wide because there is a need to change existing educational systems and curricula. NRM is always the result of interaction between three different systems — resources, resource users, and the larger geopolitical systems in which they operate — each of which needs to be understood and appreciated if intervention is to be effective.

Introduction

1

The Nature of Natural Resource Management

Natural resource management (NRM) refers to the means by which benefits are derived from natural resources. The means are diverse, both direct and indirect, and the nature of the benefit must be stated for NRM to have any meaning. Who defines the benefit — resource users or outside observers — is relevant, and it is also helpful to specify which particular resources are of interest. Seldom can the entire range of natural resources be improved or put to use in equal measure. Almost invariably NRM seeks to disadvantage some component of the biome — that which is rated undesirable or noxious in the context of prevailing management objectives.

The 'means' in pastoral areas center principally on the deployment of livestock — the species and numbers of livestock that are kept, where they are herded, and at which season of the year. Pastoralists also exercise NRM by direct intervention such as cutting, gathering, and burning. And several categories of external input can have a major impact on NRM, particularly supplementary feed and veterinary medicines, whether secured by the pastoralists themselves or provided by the State.

The 'benefit' that pastoralists seek from NRM can be expressed principally in terms of food security and the continuing ability of their livestock to contribute to their social and economic needs. Maintenance of milk and meat output are prime considerations, but the numerical size of a family livestock holding is also important. Building a sizeable holding acts as insurance against loss from drought and disease and also helps to maintain social obligations. A surplus that enables sale of animals may be an additional consideration. Most pastoralists now rely on market transactions for part of their subsistence, as well as for the 'extras' that add to the quality of life.

Other stakeholders in the pastoral system may see the benefits of NRM differently, such as emphasizing economic offtake or environmental conservation. These perspectives are no less valid, although they are secondary in the sense that the starting point for NRM intervention in pastoral areas must be from the perspectives and practices of the pastoralists (the resource users).

This chapter therefore has two objectives — to explain NRM in terms of the separate but interacting systems on which it is based, and to outline the principal instruments through which pastoralists exercise NRM.

Three systems contribute to NRM:

- the natural system, formed of the interaction among landform, climate, flora, and fauna;
- the user system, determined by how the resource users, individually and corporately, meet their biological, social, and economic needs; and
- the larger geopolitical system, comprising state policies and institutions, the external entities and forces which influence these, and all else that determines opportunities for trade and territorial expansion by resource users.[1]

Natural System

Two features of the natural system stand out as influential in NRM:

- the variety of resources that need to be managed, and

1. Employment opportunities outside of herding also lie within the larger geopolitical system, and may need to be developed in order to relieve pressure on natural resources. However, emphasis in this paper is on *direct* intervention in NRM.

- the zonal variation in climate and landform that determines the nature of resources and their response to management.

Resources to be managed vary in both their nature and utility:

Water is unique among resources because it controls the use that is made of other resources. Only where water is permanent can there be settlement or dry-season grazing. Areas without reliable wells or permanent rivers or lakes can only be used as wet-season grazing areas. The latter often must be vacated while grazing and water are still available so that the users can get back to their dry-season refuge while there is still water en route. Strategic water development can overcome such obstacles, but to introduce permanent water (if that is possible) invariably attracts new settlement and disrupts the regional pattern of resource use.

Grazing is always in the forefront of resources being managed, and often calls for additional attention to reduce over-exploitation. The need for improved management is most likely where soils are highly erodible. The key characteristics of the grazing resource are variable supply and quality associated with shifts in rainfall quantity and distribution:

- feed supply increases but quality decreases during the course of the growing season;
- feed quality also tends to improve as annual rainfall declines, especially the quality of 'standing hay' which cures better and suffers less from leaching in arid regions;
- bimodal rainfall (as in East Africa) and winter rainfall (as in North Africa) are conducive to the establishment and survival of perennial grasses under a low annual rainfall;
- perennial grasses have the advantage, relative to annuals, that they have a longer grazing season and keep more of their roots and herbage intact through long dry seasons; and
- woody vegetation has the further advantage of providing nutritious leaves and pods (browse) when grasses are dry and relatively low in nutritive value.

Woody vegetation constitutes a resource in its own right as a source of both browse and fuel. It is almost always critically important as both, although it may also need to be managed to control undue bush encroachment on grazing land.

Other plant products are also widely used, both domestically and for trade, and their sources may need to be managed to counter over-exploitation (especially sources of food and medicine, and of materials for carving or weaving).

Fauna may be similarly exploited for food or artifacts, or controlled in the case of predators. Sometimes pastoral communities are incorporated in wildlife management schemes (for example, CAMPFIRE in Zimbabwe; Box G-1), but more usually responsibility is assumed by the State or let to commercial interests.

Mineral resources are also usually removed from pastoral control, invariably when the resource is oil or a major deposit of gemstones or precious metals. Minerals of lesser value, however, are often left under local control.

Soil is managed indirectly through grazing management unless it is cropped (such as by agropastoralists). In this case, soil fertility maintenance becomes a central NRM issue.

When a resource is taken out of pastoral control, a mechanism may be established to allow local participation, but the forms of organization used for that purpose will not be the same as those adopted when communities are exercising direct management responsibility. In the latter case, the organizational form must be tailored to the resources and tasks involved. *This is possibly the most critical aspect in assisting NRM — to ensure that the membership, governance, constitution, and status of organizations with management responsibility are appropriate to the extent, characteristics, and management requirements of the resource(s) being managed.*

Zonal variation changes the nature of the resources being managed and hence the nature of NRM. In the context of NRM, the zonal influence affects management strategies more than forms of organization. The definition of a zone is complicated. Normally zones are differentiated on criteria of latitude, elevation, and rainfall, but these are not the only relevant factors. Regional hydrogeology would be invoked where occupancy depends on the presence or absence of groundwater. And although edaphic criteria are more suited to zonal subdivision, floodplains of regional significance might well be demarcated at zonal level (for example, the inner delta of the Niger River in Mali).

The primary criterion in the tropics is relative aridity. This prescribes at least five zones, ranging from true desert to the dry subhumid zone. Deserts only contribute grazing and wildlife resources if they support some plant life, but even barren lands can contribute useful minerals. The dry subhumid zone is also of limited use to pastoralists — once it was widely prized as dry-season or drought reserve, but now it is usually occupied by agriculturists and accessible only through trading relationships. In between these extremes are:

- the semi-desert, where rainfall is very sparse, vegetation is ephemeral or confined to depressions and water courses, and livestock (camels and goats) subsist largely on browse;
- the arid zone with monomodal rainfall, where conditions are less harsh than the semi-desert but where cattle have difficulty surviving once the annual grass cover begins to disintegrate;
- the arid zone with bimodal rainfall, which has the potential to support perennial grasses (along with thorn-bush), and where most livestock species fare well in favorable years; and
- the semi-arid zone, where rainfall permits more intensive resource use, including rainfed cropping (although *reliable* cropping and sedentary herding require a wetter zone).

Rainfall data are omitted from this list because mean annual rainfall (the figure most commonly cited) is illusory. Appending coefficients of variation helps, but not greatly unless the average is truly the norm, which usually it is not (Tyson 1986). The pattern of rainfall over much of Africa this century has shown alternating wetter and drier periods, each lasting a decade or two. A 'dry' period has about 70 percent of years below the long-term average, and a 'wet' period a comparable over-representation of above-average years.

These effects are most profound in or fringing the semi-desert, where perennial grasses and surface water are relatively abundant during the wetter periods but are absent or unreliable during the drier periods. NRM strategies must adapt with each change, more so under monomodal than bimodal rainfall because drought expectations are different. In Kenya, for example, bimodal rainfall in the northeast produces dry seasons lasting only three to nine months, with an eighteen-month drought occurring every ten years or so, whereas rain failure in the monomodal areas of the northwest frequently produces droughts lasting two to four years.

User System

The user system comprises the resource users, their property (including livestock), and their own forms of social-territorial organization. If more than one social group is involved, then the interaction among groups is also part of the user system. The characteristics most fundamental to NRM are:

- demographic trends, covering population, wealth categories, customary demographic controls, and projected changes;
- the communities involved, differentiated by ethnicity, size, resource interests, and relationship one to the other;
- their production systems, defined in terms of livestock species and the uses made of these and other resources;
- relationships with neighboring agriculturists, outside political powers, and the market economy; and
- existing social-territorial organization, detailing social structure and systems of property rights and their significance in NRM.

Establishing these characteristics usually requires detailed field study. There are no social databases comparable with the resource maps that help categorize the natural system. The last three characteristics in the immediately preceding list serve to show how present NRM relates to the goals and value systems of resource users and to their existing social institutions. This, in turn, shows where improvement can be made. Consideration of demographic factors (plus geopolitical issues) then moves the focus from priorities in NRM to aspects of practicality and sustainability.

Demography is a key factor because population pressure is both a cause of mismanagement and a constraint to remedial action. Pastoral societies have a number of strategies of their own for adapting to population growth. They cannot intensify like agricultural societies, but they can control population growth by controlling fertility. They can subsequently adapt by trading more, adopting agropastoral practices, and diversifying into other activities (including paid employment). Three aspects warrant elaboration:

Population pressure is a concept that takes substantial form only when related to the support capacity of the space or the resource being utilized and an expected standard of living. It is not a straightforward statistic like population density, and its significance in NRM also depends on the manner in which resources are being used. An agropastoral system can sustain more people per unit area than a pastoral system, while a trade-based pastoral system can support more than a system requiring a dietary milk supply year-round.

Wealth differentiation results in households of different stockwealth accessing different resources. It is therefore important to extend analysis of user systems down to the household level (see Lane 1996), and to establish causes as well as occurrences of stockwealth variation. A high proportion of chronically poor families is always a complication and may preclude sustainable NRM if the poverty trend is increasing. A high proportion of livestock owned by non-residents — as a result of absentees investing their wealth in livestock in their home areas — will also radically affect strategies for NRM.

Overlap among pastoral groups becomes especially troublesome if there is group competitiveness and antagonism. Events in Rwanda and Somalia, and 'ethnic cleansing' in Europe itself, begin to bring home to Western observers the potential persistence and destructiveness of ethnic antagonism. The prevention of armed conflict is now part of the role of the larger geopolitical system, but it is well to remember that territorial expansion is a well-tested NRM strategy, and still the surest way of coping with drought.

Other features of the user system that influence NRM are types of animals kept, uses to which they are put, and aspects of mobility, resource tenure, and decisionmaking that are elaborated later in this chapter. Additionally, societal goals and value systems need to be accurately interpreted because they are the foundation on which present NRM practices are based. One of commonest errors in development planning has been to impute alien value systems.

Geopolitical and Macroeconomic Framework

The geopolitical environment, formed of State and other external influences, is the most diffuse of the parent systems of NRM. It is as influential as the other two systems and possibly dominates in determining the effectiveness of present NRM and the options for improving present practices.

The State can support pastoral development in several ways. The usual avenues are through education, public and animal health services, new roads and water supplies, and research. Also important to NRM is the institutional framework that governments provide in terms of land policy, the legal code, and security against unwanted incursion. Seldom, however, is the requisite help forthcoming. Only in a handful of countries, out of about 40 where extensive pastoralism is practiced, do pastoralists command automatic political or economic priority. And even then, ethnic rivalry and a weak economy may intervene. Elsewhere, in most countries, support to pastoralism is low on the political agenda.

Where pastoralists have minority status, they inevitably become marginalized and face the threat of acquisition of resources for majority interests. Sectoral planning increases that threat because pastoral economies are perceived as having no competitive edge — they are treated either as a low-grade extension of the livestock economy or as a risky partner in wildlife-based tourism. And when the commercial private sector is encouraged to participate in service delivery, pastoral areas tend to lose out because there are quicker returns to be made elsewhere.

External assistance usually determines the extent to which the State actually intervenes in pastoral areas. Until recently, the resulting projects have tended to emphasize infrastructure and technical services and pay less attention to the institutional framework. Rarely has there been support for review or reform of land and corporate law for the benefit of pastoral tenure and pastoral institutions. Moreover, because aid has flowed to central governments, little assistance has been given to local government structures to assist them in supporting community-based development. Sometimes NGOs have stepped in to fill this gap, but NGOs provide only stop-gap solutions.

Attitudinal problems abound in the geopolitical system:

Sedentarization is often State policy, with emphasis placed on ease of administration and service-delivery, rather than on maintaining mobility and the integrity of separate wet-season and dry-season grazing areas. This happens even when politicians and technicians are of pastoral origin because of peer pressure and the conditioning received during their formal education. It happens less in local government circles than in central government — but local government does not hold much power and responsibility. If sedentarization is combined with sectoral (livestock) development, then particularly poor and damaging projects can result.

Participatory planning is now the rule rather than the exception, but decisions on how to apply planning data remain as the preserve of administrators and planners. There is still a lot of lip service to participation and not much attention to ensuring that local groups have forms of organization and counseling appropriate to the roles expected of them.

Overgeneralization applies not just in participation but also in most literature supportive of pastoralism. Until supporters and planners acknowledge the big differences that small changes in environment and resource-use make, there is not much hope for better development projects.

These attitude problems are not confined to the geopolitical system. There is increasing demand for sedentarization among pastoralists too. No doubt pastoralists who want to settle should be enabled to settle, but only if it is understood that neither money nor technology can produce rain or unlimited groundwater, and that their livestock (currently their larder) must continue to move, or die. For the sake of all stakeholders in pastoral society and outside, there is need for re-education and change in the curricula responsible for present attitudes.

Implications for Customary NRM

Present practices in NRM are shaped by all three of the systems just described. Although the influence of the three systems described can be separated — as happens when scientists and planners apply their individual perspectives — this always carries risk of misrepresentation. Usually it is the *interactions* among systems that are most influential.

It is to minimize misrepresentation that the next chapter argues strongly for the categorization of pastoral systems prior to intervention. This serves to prescribe the situation being assisted in terms of attributes deriving from each parent system. Task managers will not find much guidance outside this document, however, about how development requirements vary according to pastoral system. Most writings on pastoralism present an homogenized view of what pastoralism entails, even when dealing with specifics such as resource tenure.

Pastoral resource tenure has attracted several reviews as part of a movement that seeks more sympathetic treatment of common property regimes (Bromley 1992, Birgegärd 1993, Lane and Moorehead 1994). However, the discussion is still at the stage of arguing the merits of alternative theories of tenure and of mobility as a resource-use strategy, rather than adapting theory to practice. There is still confusion between access and tenure, and still a tendency to lump all natural resources rather than to view them, as pastoralists do, as separate resources with temporal as well as spatial identity. This section focuses on resource tenure, and further background information on access and mobility is included in Annex A.

Grazing has ambiguous property status. True nomads grazing among more sedentary people may never translate access into tenure, while pastoral societies with restricted territory can be excessively possessive of grazing. Most possessive will be the level of society responsible for maintaining boundaries. It is usual for societal property regimes to be more effective in controlling use by outsiders than grazing pressure by group numbers. And in addition to societal variation, the property status of grazing varies among localities and seasons.

For most purposes, however, grazing can be regarded as common property over which preferential rights can be established. It is quite usual for the pastoral group most closely associated with an area (whether by residence, delegated custodial responsibility, or historical association) to have preferential use of that area. This does not preclude other users, but they are expected to ask beforehand and to provide reciprocal grazing if required. Some tenure systems permit groups or individuals to demarcate localities for exclusive use, but such areas are usually small relative to the total area of grazing available.

Preferential and exclusive rights apply particularly to dry-season grazing. Wet-season grazing is nearly always unattached common property. The enhanced property status of dry-season grazing is locational as well as seasonal because dry-season grazing has, by definition, to be accessible from permanent water.

Water varies from being an open access resource to private property. At risk of overgeneralizing, water is:

- an open access resource if it is the direct product of rainfall and not entrained in streamflow or contained in storage to which property rights are attached;
- unattached common property if it is in a seasonal water course or natural surface storage in a wet-season grazing area that has no other use;
- attached common property (conferring preferential rights to one group) if it is in or under a water course that is exploited for irrigation or as a dry-season water supply; or if it is contained in storage that is identified with a specific kinship group or situated in an area to which preferential grazing rights are attached; or
- private property if it is already diverted or otherwise removed from streamflow, or if it is contained in a well or other storage identified with an ancestral or living person.

When water is private property, it may still be made available for wider use, but usually on the basis of kinship and in return for input to the maintenance of the well or storage facility concerned.

Other resources are less vital to pastoralism than water and grazing, and attract property status according to use and user:

Browse is normally regarded as an adjunct to grazing, although special arrangements may enable access to browse plants of particular value (for example, stands of salt bush that are a unique source of minerals in the region).

Other plant products (charcoal, edible parts, fiber, gums and resins, medicines, poles, etc.) are also secured as though grazing were being accessed, unless extraction is by specialists with separately established rights or unless trees of specific property value are involved.

Individual trees are sometimes privately owned, but usually through inheritance because the practice of attributing property value to trees is declining. It can happen that trees so owned are now under the territorial control of other resource users, which makes the value of ownership dependent on how the present regime treats such rights. Rights to access trees to hang beehives and collect honey can also be assigned and inherited.

Wildlife is either an open access resource or attached common property, unless subject to royal prerogative or appropriated by the State. Attached status usually applies selectively and temporarily, when species of value (such as wild or feral asses) are present in the area controlled by the resource users.

Mineral resources are now mainly under State control so that customary management has little relevance. Where salt and smelting materials are still extracted customarily, it is often by specialists exercising inherited rights. Minerals not covered by old rights or State controls are an open access resource. Salt licks often carry property rights similar to salt bushes (see *Browse*, above).

Part One
Preparing for Project Intervention

The guidelines presented in this paper fall in two categories. This first section establishes a framework of effective support for pastoralism, while the second covers individual project components.

Effective intervention in pastoral situations has four basic requirements:

- to differentiate the type of pastoral system(s) being assisted, including elements of mobility, type of livestock, involvement in other enterprises, and relation to external economies;
- to assess population pressure and societal coping mechanisms, in order to establish whether present population pressure represents a serious impediment to progress;
- to ensure that enabling policies and infrastructure are in place to support pastoral development; and
- to ensure a project design that combines participation, flexibility, and the prospect of sustainability.

The next five chapters address these issues. Chapter 2 categorizes the main pastoral systems and shows how differences in their characteristics change priorities in NRM and project inputs. Chapter 3 pursues the same theme in problem situations where there are indications of population pressure and overgrazing. The following two chapters examine key features of the geopolitical system and how to create an enabling environment for effective intervention. Macroeconomic issues are considered in Chapter 4, and institutional ones in Chapter 5. Chapter 6 considers how to incorporate notions such as participation and sustainability in project designs.

2

Categorizing Pastoral Systems

Pastoral systems are systems of natural resource use in which free-ranging or grass-fed animals are the principal means of exploiting the resource. The animals can be farmed or ranched if kept on private land, herded or shepherded when kept on communal land, or hunted or mustered in the case of wild or feral species.

Pastoralists engage mostly in herding.[1] Pastoral groups with land title can be said to be ranching, although often they retain herding practices. Some agropastoralists now 'farm' livestock by stall feeding cut grass, but this practice too is localized (although less so in Asia than in Africa). Hunting, which once was quite widespread, is now usually discouraged, while mustering in Asia and Africa is restricted mainly to feral horses and donkeys.

While these broad features of management begin to differentiate pastoral systems, they are insufficient in themselves to define development needs. For that purpose, it is necessary to consider a wider range of attributes (Table 2-1). Some of the characteristics listed are mutually exclusive or are very local, but there are still about twenty quite distinct pastoral situations that planners are likely to encounter in the arid areas of Africa and Asia. Some of these are listed in Tables 2-2 and 2-3 to show how they vary in the assistance that they are likely to require. The main categories are also described in more detail in Annex D.

The remainder of this chapter elaborates four attributes that are particularly important in determining development inputs and the chances of success — mobility, the nature of the livestock resource, economic orientation, and social-territorial organization.

1. Shepherding is differentiated in this document only when sheep clearly dominate the pastoral system.

Mobility

Mobility is a common feature of pastoralism, but varies greatly in its practice. It varies in range and seasonality and in the linkage that is maintained between people and livestock. Nomads move as whole families with all their livestock, while pastoralists who engage in transhumance (following set geographic and seasonal routes) or who move less frequently or in a more constrained fashion may delegate just a few herders to accompany the livestock. In the latter case, it is quite usual for different classes of livestock to be herded according to different patterns, with some animals retained close to the family homestead.

Mobile pastoral systems are a natural adaptation to the unreliable supply of feed and water in arid environments. Even under less arid conditions, mobility can be beneficial in raising livestock productivity, but under severely arid conditions, mobility is a matter of life or death. As a general rule, any action that curtails transhumance or otherwise restricts mobility is to the detriment of pastoral production. At the same time, movement patterns arise in several ways. Some are imposed by climatic or other forces, while others reflect the predilections and choices of the pastoralists. Intervention must be preceded by analysis of the rationale for movement (Annex A).

Semi-sedentary systems often arise from loss of mobility. This may be because (a) pastoral territory is being lost to other uses; (b) the State has a deliberate policy of sedentarization; (c) people prefer, and new wells allow, fewer moves; or (d) loss of stockwealth enforces reduced mobility. In the last case, the resulting semi-sedentary system may occur *within* a mobile system, practiced by families with fewer and less mobile livestock than those who move freely.

But there are also systems that are semi-sedentary because there is no need to be mobile. Usually this situation arises in zones that are subhumid or wetter, but it also occurs in arid zones where there is a reliable source of dry-season feed from irrigation, a processing industry, or marine resources. Dried sardines have supported pastoral systems bordering the Arabian sea for many generations. Distinguishing the causes of reduced mobility is essential for effective intervention in semi-sedentary systems.

In general, development should aim to assist mobility, but where there is no need to be mobile, or pastoralists are caught in a poverty trap, semi-sedentary systems warrant support.

Livestock Resource

The livestock species associated with a pastoral system are determined both by the environment (climate, terrain, disease hazard, and type of feed) and by the functions required of stock (in terms of providing milk, contributing draught power, building social relationships, etc.). Seldom is there just one species, and there may be four or five, each fulfilling a different but complementary function. Seldom, too, is any species reared solely for sale. Usually the animals that are sold are of the type most abundant or

Table 2-1. Criteria for Differentiating Pastoral Systems

Attribute	*Subset*	*Feature*
Ecology	Locational	Temperate or tropical (also sometimes seaboard and upland) [a]
	Pluvial [b]	Desert, dryland, or (sub) humid
Economy	Animal-based subsistence	Meat-oriented or milk-oriented
	Diversified	Agropastoral, investment-based, or trade-dependent (also sometimes multiple-use) [c]
Livestock	Bovine	*Bos indicus* (zebu) or other cattle (plus yak in E. Asia)
	Camelid	Bactrian camel or dromedary [d]
	(Other)	Horse, mule, donkey, sheep, or goat
Territoriality	Mobile	Nomadic, peripatetic [e], or transhumant
	Restricted (semi-sedentary)	Centrally planned, naturally restricted, ranch-bound [f], or resource-poor

a. Upland and seaboard need citing where they shape the system climatically or provide fish as stockfeed.

b. Because mobile pastoral systems cross zonal boundaries, only general terms are needed to indicate adaptation to aridity.

c. Multiple-use can be specified if income derives specifically from wildlife (for example, through tourism, game ranching, or sale of gums and resins).

d. If Latin America were included, the domesticated alpaca and llama and the wild vicuna would need to be added.

e. Peripatetic describes mobility that is more restricted than nomadic and less regular than transhumance.

f. Ranching takes many forms — imposed forms can be grouped with centrally-planned enterprises as institutionalized.

Table 2-2. Mobile Pastoral Systems

Category	*Economic orientation* [a]	*Priority for intervention* [b]
Desert camel herding		
Tropical (e.g., North and East Africa/Arabia)	Milk-subsistent (now sometimes motorized in oil-rich states), based on dromedary (milk, transport) and goat (meat, milk, barter)	NRM: drought strategy, well improvement, browse management, oasis agriculture where relevant. Trade: possibly offtake of small stock
Temperate (e.g., Mongolia)	Trade-dependent, based on Bactrian camel (hair, transport) and sheep (wool, meat, milk, sale)	NRM: drought strategy, well improvement, browse management, oasis agriculture where relevant. Trade: wool/hair, live sheep
Dryland shepherding (e.g., North Africa/ West Asia)	Meat-subsistent and trade-dependent, based on sheep (meat, milk, wool, sale), goat (hair, milk, meat), and horse (transport)	NRM: strategic water development, some range seeding. Trade: mainly live sheep plus hair and wool, including breeding for commodities, e.g., pelts
Dryland and transzonal cattle herding		
Peripatetic (e.g., East Africa)	Milk-subsistent, based on zebu cattle (milk, sometimes blood), sheep and goat (meat, barter, milk), and donkey (transport)	NRM: key resource improvement, strategic water development, drought strategy. Trade: possibly offtake of small stock/immatures, gums and resins
Transzonal transhumance (e.g., West Africa)	Trade-dependent, based on zebu cattle (milk, sale and barter/sale for grain) and sheep and goat (meat, barter, milk)	NRM: as appropriate to zone, fire and possibly seeding in subhumid zone, plus enhancing crop-livestock complementarity. Trade: build on existing trade
Agropastoral transhumance (e.g., Africa/India)	From milk and crop-subsistent to trade-dependent, utilizing zebu cattle (milk, tillage, manure, sometimes sale), goat and/or sheep (meat, barter, milk)	NRM: as above plus cropland improvement. Trade: as above, possibly with stall feeding

Note: See Annex D for further description of these and allied systems.

a. Species and functions of livestock that characterize the system are indicated, with the principal species underlined.

b. Inputs to all mobile systems should maintain or strengthen existing territoriality, support herder organizations (especially in conflict resolution and welfare support); provide mobile services and strategic service centers.

most readily replaced.[2] Moreover, most species have the potential to fulfill several functions (Table 2-4), so the actual function to which a species is put depends on which other species are present. This varies not just among pastoral systems, but also among households according to their status (including their stage in the cycle of household formation, maturation, and dissolution).

The consequence of these factors is that development cannot be targeted to improve or commercialize one species without first assessing the impact on other species and on the integrity

2. Unless, of course, a large sum of money needs to be realized from a sale, when it may be better to sell one camel rather than a whole flock of goats. Managing a livestock holding is much like managing an investment portfolio: what one sells is determined by how much money needs to be realized and by one's judgment on how each holding is likely to perform if held.

Table 2-3. Semi-Sedentary Pastoral Systems

Category	*Economic orientation* [a]	*Priority for intervention*
Resource-poor goat herding (e.g., Africa/India)	Milk and meat-subsistent, based on <u>goat</u> (milk, meat, barter), sheep, sometimes with a house-cow and/or donkey	NRM: browse management plus inputs to goat husbandry Trade: possibly gums and resins
Localized cattle herding	Several distinct systems (peri-urban, off-farm, seaboard, and upland) occur in Africa and Asia, usually milk-subsistent or trade-oriented, based on small herds (sometimes camels in Yemen) and often some small stock	Interventions depend on local circumstances, focusing on maintaining or improving dry-season feed supplies and ensuring adequate health services
Ranching		
Group ranching [b] (locally in Africa)	From milk-subsistent to trade-dependent, based on <u>zebu cattle</u> (milk, sale), sometimes with sheep and goat (meat) and wildlife (revenue from tourism	NRM: grazing management, key resource improvement, agistment [c], intensifying management and crop-livestock complementarity as group areas subdivide Trade: build on existing trade, diversify as possible
Commercial ranching (e.g., southern and East Africa, locally elsewhere)	Investment-based and trade-dependent, based on <u>zebu and crossbred cattle</u> (meat and stock sales), occasionally other species and enterprises	NRM: as above; ensuring social and ecological feasibility and legal instruments prior to ranch development Trade: pursue new market opportunities

Note: All restricted systems encourage private sector inputs (for example, animal health services) and diversified production.

See Annex D for further description of these and allied systems.

a. Species and functions of livestock that characterize the system are indicated, with the principal species underlined. Although only commercial ranching is listed as investment-based, this orientation is increasingly evident in parallel with subsistence and trade-dependent herding (Annex D)

b. Group ranches arise when customary land rights are adjudicated in favor of pastoral groups, with subsequent allocation of freehold group title. Other types of institutionalized ranching occur in central Asia, initially centrally planned and now being liberalized.

c. The leasing (or reciprocal exchange) of grazing at times of seasonal need.

Table 2-4. Uses of Livestock in Pastoral Systems

	Environmental adaptability		Feed intake		Utility (customary use × yield or quality)						
Species	*Desert*	*Dryland*	*Grass*	*Browse*	*Milk*	*Meat*	*Wool/hair*	*Hides/skins*	*Transport*	*Draught*	*Manure* [a]
Camel	+++	++	++	+++	+++	+	++	++	+++	+	+
Sheep	+	+++	+++		+	+++	+++	++			+
Cattle [b]	+	+++	+++	+	+++	++		+++		+++	++
Goat	++	+++	++	+++	+	+++	++	+++			+
Horse		++	+++		++	+		++	+++	++	+
Donkey [c]	++	++	++	+	+			+	++	++	+

Note: Principal species only (excluding pigs, poultry, and bovids and camelids of local importance) and generalized across a wide spectrum of pastoral systems. In practice, 'draught' applies mainly to agropastoral systems, and the ratings shown for other uses vary among individual systems. The ratings for 'meat' recognize that large stock are usually too valuable for regular slaughter, while small stock are often kept specifically for that purpose. In several systems, sheep and goats deserve a ++ rating as a source of milk.

a. Manure can be used either as fuel (where fuelwood is in short supply) or, in agropastoral systems, as fertilizer.

b. Cattle are also the principal source of blood where this is a dietary supplement, and of horn.

c. In its water metabolism, the donkey is as well adapted as the camel to desert conditions. It can also cope better than the camel with rocky terrain, but has less load-bearing stamina. Donkeys can also graze spiky herbage (e.g., *Sporobolus spicatus*), which other livestock avoid.

of the pastoral system, considering not just the overall system, but also at the household level.

Economic Orientation

Most pastoralists drink milk, eat meat, and collect or purchase other food items. There are, however, major differences among pastoral societies, and hence among pastoral systems, in how the diet is constituted and how much food (among other commodities) is purchased. These differences do not necessarily show up strongly in the contribution that milk, meat, and grain make to energy intake, but a preference for milk or meat greatly influences how NRM is practiced, even if that commodity contributes only a fraction of total energy intake.

Milk-subsistent systems have milk as the staple. Purchases are confined to additives, or to times when milk and other livestock products and bush foods offer a seriously inadequate diet. There is no felt need to eat grain foods on a daily basis, and certainly no desire to sell or barter livestock to secure grain regularly by the sackful. Livestock may be sold for other reasons, such as when a specific need for cash arises, but they are too valuable to be squandered. Therefore, positive supply responses cannot be assumed. In Africa, this situation is more common in East Africa than elsewhere. The implications are seen not just in the seasonality and categories of livestock offered for sale, but also in NRM. More so than in other pastoral systems, pastoralists who are milk-subsistent have to make the very best use of grazing and browse on offer in order to keep animals in milk year round as nearly as possible.

Meat-subsistent systems have a greater consumptive demand for meat. Meat intake may be small relative to other foods, but still exceeds what is normal in milk-oriented systems. The meat comes principally from small ruminants, especially sheep, and is commonly eaten with bread. If the grain for the bread is not home-grown, it must be purchased, which makes many meat-subsistent systems also trade-dependent. Milk is an integral part of the diet (often taken as yogurt or cheese), but in quantities that the sheep flock can provide. Because the system centers on sheep rather than cattle or camels, NRM reflects the dietary preferences and grazing behavior of sheep.

Trade-dependent systems arise whenever there is reliance on purchased items. Grain is not the only purchased item, but is the most common. Most pastoral systems of West Africa are trade-dependent because of a preference for grain foods. In that region, customary transhumance patterns assist procurement by incorporating annual stopovers in grain-producing areas. In other systems where trade is not necessarily associated with grain purchase, movement patterns pivot around bringing livestock to market where and when prices are highest (for example, at times of Id festivals in the Muslim calendar). Typically, with increasing prices, supplies will increase.

Investment-based systems represent a different situation, where people in business or paid employment choose to invest capital in livestock in a pastoral area to which they have right of access. Such systems are now widespread. These livestock usually benefit herders in the area, but their management does not necessarily conform to customary NRM, and they add significantly to grazing pressure. Moreover, accepting animals from an absentee owner is often the start of the slippery slope out of pastoralism.

Yet another situation arises when money is invested in a vehicle to transport water and/or livestock to areas (usually remote desert) with unused feed resources. This category and some others mentioned above are described in more detail in Annex D.

These distinctions are often overlooked when planning marketing infrastructure to serve or tap pastoral areas. Planners tend to see livestock only as a tradable commodity to be valued in terms of retail value. Developmentally, however, it is vital to know whether the animal on offer is being sold because (a) it is surplus to subsistence needs, (b) the need for cash overrides but does not seriously impair family security, or (c) it is an act of last resort which threatens or eliminates the viability of that family as pastoralists. The cash value that a pastoral family (or pastoral society at large) places on livestock of any particular category reflects the value of that type of animal within the pastoral economy, which may bear no relation to market value.

Social-Territorial Organization

Social-territorial organization is a convenient way to describe the matrix formed of societal structures and territoriality. This multi-dimensional matrix has strands contributed by customary ethics, environmental necessity, and accumulated management experience, and is also shaped by external pressures and changing demography and expectations within society. Mobility is one expression of social-territorial organization, and is a particularly important aspect. But how societies are structured, and the mechanisms by which they control access to resources, are also important features that need to be differentiated. Of particular significance is the effectiveness of pastoral institutions in handling problem situations. The follow features are especially important:

Social differentiation needs to be examined in depth in order to clarify who in society controls decisionmaking and how benefits are likely to be distributed. It is not sufficient just to establish the framework for decisionmaking. Caste and gender may be as important as wealth in determining who may hold, inherit, and dispose of property rights, and who can take up or benefit from new opportunities provided through project intervention.

Flexibility is an important attribute in social structures and family responses. Few pastoral systems are static. Most systems face a dynamic environment and change in form as they respond to new challenges. People move in and out of systems, systems themselves adapt and evolve, and adaptation brings changes in the relationships among systems and with other economies. Although we know too little about systems dynamics to always be able to predict capacities for change, it is possible to focus on a few key areas, as highlighted below.

Adaptations to increasing population pressure warrant particular attention. Most pastoral systems have developed procedures to cope with population increases, which tend to be distinct from those applying to agricultural systems. Pastoralists cannot readily intensify production, but typically they show lower fertility than farmers, and also show considerable flexibility in moving out of and back into pastoral activity as circumstances require. Establishing how particular societies and families cope is critical to the definition and delivery of effective development assistance.

Interactions between agriculture and pastoralism are a specific area of interest. One of the most common ways to cope with increasing population pressure is to increase cropping or relationships with neighboring agriculturists. Obviously, this is easier to accomplish in wetter areas and in systems already accustomed to transhumance than in situations where people are tied exclusively to arid lands. The latter are more likely to turn to (temporary) paid employment than to agriculture.

Relations with market systems are equally important. It is still relevant to differentiate systems according to their relative trade orientation, but few pastoral systems are now independent of markets. For the great majority of contemporary pastoral groups, efficient markets are a key determinant of well-being. Ultimately, marketing hinges on *family* needs and circumstances — but how *groups* handle trading relations is an important attribute when deciding how to deliver development support.

Drought strategies are a practical expression of the flexibility and organizational 'competence' of pastoral systems. In many respects, these strategies are the summation of all of the foregoing. Although families have individual responses, the organization of responses at the group level — not least through internal welfare support mechanisms — is very informative of the ability of pastoral systems to cope. Drought management is covered in its own chapter in Part Two of this document.

Conflict and conflict resolution have dimensions 9extending well beyond internal social-territorial organization. Far too many pastoral systems are currently entrained in far-ranging ethnic and political conflicts. How societies resolve their internal affairs is no measure of how well they can cope with external conflicts, but establishing the effectiveness of customary institutions in conflict resolution is still critical in development planning. Indeed, one powerful justification for renewed interest in pastoral development is the potential to use customary institutions to help defuse conflicts that have causes in natural resource competition.

3

Assessing Population Pressure and Overgrazing

Population pressure must be assessed in concert with the categorization of pastoral systems. Invariably, population pressure creates situations calling for intervention, but the more acute the pressure on resources, the more difficult it is to intervene successfully. It is a critical aspect of project planning therefore to identify when the pressure on resources becomes a major impediment to progress.

Relevant Perspectives

Until recently, there was consensus that population pressure was crippling the ability of pastoral areas to support future generations. Then came the observation that despite decades of apparent misuse, the support capacity of the land seems not to have declined. And from the marriage of that observation with the notion of non-equilibrium environments as having a variable but resilient grass cover, came the conclusion that heavy utilization of arid rangeland by pastoralists is not to be regarded as overgrazing. This line of thinking is elaborated in Annex C.

In reality these observations depend on the perception of overgrazing. The above sequence draws heavily on the thesis that pastoral societies must expect to 'bust' as well as 'boom'. Similarly, some argue that it does not matter if some areas are degraded and eroded, if run-off soil and water collect elsewhere in the landscape. This is a relatively narrow perspective that gives no weight to human suffering that occurs at the family level as populations contract and expand, and it discounts the wider implications of soil and water loss from degraded areas. In a development context, sharper ecological and humanitarian perspectives are needed.

An ecological perspective needs to embrace entire systems and comprehend time as an ecological factor. Over time, each rangeland site has the possibility of at least three ecological states:

- the natural state represents what the site would be without human influence, given the climatic and geomorphic conditions (including variability) that prevail;
- the derived state represents the condition to which the site can be brought by concerted human use, without special effort but merely through consumptive utilization; and
- the deformed state represents the condition that can arise from more acute interference, without waiting for geological time to affect that extent of transformation.

The same basic ecological processes are at work across this spectrum of ecological states. However, the same processes produce different results (different states within states) when working on different life forms and different availability of life-support materials (in terms of water, O_2-CO_2, solar energy, and minerals). There is, moreover, interaction between processes and any new baseline situation that forms, so very different ecologies can arise from small changes in one factor.

No one state (or state within a state) can be said to be more productive than another unless a measure of productivity is specified. According to the thinking outlined in Annex C, the derived state can be more productive in terms of a pastoral survival strategy than the natural state. But it is not yet clear whether that strategy, projected through time, will lead to a less productive state, or whether a survival strategy will continue to satisfy the resource users. Wildlife and ranch managers usually prefer a strategy that stays closer to the natural state.

A humanitarian perspective cannot readily accept a resource management strategy that causes families to lose livestock and livelihoods, and then to depend on famine relief for the years that it takes to rebuild herds. At the very least, the present and projected extent of hardship must be quantified.

This requires project planning to be accompanied by an explicit statement of the standard of living that is to be accepted or achieved. Without such a statement, other stated objectives remain nebulous. The statement cannot realistically be tightly quantified, but should specify how different social strata will be affected and who is expected to require famine relief.

References to famine relief should not be taken as indicating that pastoral systems have an unusually high proportion of destitute people. Indeed, unless catastrophe strikes a whole society, there are better social support mechanisms among pastoralists than in most other societies. Sharing is a means of building and consolidating social relationships, and when milk is abundant, it is available in quantities that can be shared without cost to the herders or the young stock. Carcasses too often yield more meat than a family needs, although sharing meat stands in danger of being non-sustainable unless other families reciprocate.

Indicators of pressure on resources fall into three categories:

Societal indicators lie mainly in the effectiveness of welfare support mechanisms. These mechanisms are needed because of the nature of the environment in which pastoralism operates, and are not in themselves an indicator of population pressure. However, increasing use of the mechanisms or, worse still, their breakdown and replacement by famine relief, demonstrate population pressure. Of course welfare support mechanisms also decline when social values decay, but discussion with a community will serve to place what is observed in its wider context.

Household indicators lie principally in stockwealth — the proportion of families whose stockwealth is inadequate (in numbers and composition) for family security. What is 'adequate' must be established for each pastoral system and environment. There is little point in seeking to express family circumstances in monetary terms — the value of the staples of a subsistence pastoral economy fluctuates seasonally and depends ultimately on the value that is placed on life and social cohesion. More useful as quantitative support are data on demographic trends.

As noted earlier, demography is a potent force in determining the feasibility of sustainable NRM. Although pastoral societies have a variety of strategies to cope with increasing population pressure, there is always a limit to the population that an area can sustain when each family is seeking to build or maintain a subsistence herd.

Environmental indicators are the most difficult to use. Even soil erosion — a fairly obvious environmental loss — must be considered in the context of the total landscape and the resource use of the site. Loss of soil and biodiversity must also be related to a broader issue — whether the area is to be invested with a value beyond that pertaining to the present pastoral system. Some ecologists and resource managers prefer, as noted above, to look beyond present land use and see value in keeping land close to its natural state, thereby reserving options for the future. More detail on environmental indicators is included in the following section.

Zonal Considerations

Indicators of population pressure by zone are summarized in Table 3-1. The zones are those introduced in Chapter 1, and are tropical. The criteria noted for the semi-arid zone, however, also apply to zones outside the tropics.

Key Resource Areas

Key resource areas are mentioned in Table 3-1 in the context of the semi-arid zone. They occur in all zones, however, and increase in importance as conditions become more arid. They fall into several categories:

- floodplains and other major dry-season grazing areas;
- bottomlands that receive run-off and are therefore better vegetated than the rest of the landscape;
- localities with resources of specific value, either filling a nutritional gap or contributing otherwise to the economy (such as salt licks); and
- drought refuges lying outside normal grazing orbits.

The value of these areas is partly locational, but also lies in the presence of particular plant species. If those plants are destroyed, the value of the site is significantly impaired. In non-key areas, the composition of the vegetation matters less. Consequently, it is in key resource areas that overgrazing, as measured floristically, must be taken most seriously.

Overgrazing is best regarded as a condition where grazing intensity is causing the vegetation to deviate markedly from that preferred for the site and management system concerned. The basic issues are how 'preferred vegetation' is to be defined, and what constitutes marked deviation.

Preferred vegetation must be related to how the vegetation will be used, bearing in mind that different stakeholders have different views. Once those conflicts are resolved, 'preferred vegetation' is best defined by components of the current plant cover, with a view to managing what is there and favoring the most useful components (including browse). A definition based on an assumption of what plant species

Table 3-1. Indicators of Population Pressure by Climatic Zone

Zone	*Occurrence of over-population and degradation*	*Environmental indicators*	*Other indicators*
Semi-desert	Usually only around permanent water and oases	Death of browse, dune encroachment	Mortality and weight-for-age. Other indicators are unreliable, for example, often males work outside the system even in normal times.
Arid zone (for agropastoral systems, see also semi-arid social indicators)	Overall not bad, but accentuating where borehole water leads to year-round occupancy	Useful browse giving way to useless weed species, aggravated soil erosion, average rainfall years referred to as drought years	Breakdown of societal welfare support system, declining stockwealth, dependence on bush foods outside drought years (most significant in bimodal rainfall areas, where biodiversity is greater)
Semi-arid zone	Rather widespread	Extensive and worsening soil erosion (as shown by appearance of bedrock and plants on pedestals), loss of biodiversity in perennial plants, increase in non-viable cropping and marked deviation from the preferred vegetation in key resource areas	As above, especially increasing dependency on goats and famine relief (although the latter can be confounded by outside remittances), inadequate numbers and performance of draught animals in agropastoral systems

might be encouraged to return to the site is usually too hypothetical to impose on subsistence pastoralists.

Marked deviation must be related to the fluctuations in plant cover that are expected from annual and periodic variations in rainfall. Different standards may need to be set for wetter and drier cycles. In the example cited in Chapter 1 for the semi-desert, annual grasses characterize dry cycles, and perennials (which bind the soil better, start growth earlier when rains arrive, and have a longer growing season) persist only during wet cycles. In other zones, other vegetation successions are to be expected, so local knowledge is essential if standards are to be set to match cyclic variation.

Managing key resource sites lies at the heart of pastoral NRM. Management options are described in Chapter 10, but it is necessary to appreciate at the outset that these areas require individual attention. Bush fires may need to be controlled or reintroduced, and rotational grazing may have a role to improve or restore perennial grass. Certainly it makes sense to focus research on these areas where there are best prospects for improvement. One of the main contributions of rethinking range ecology (Annex C) is that attention has been focused on the needs and potential of key resource areas.

On the other hand, the overall conclusion that pastoralists do not overgraze their land is too sweeping for practical purposes. This thesis assumes mobility and a management strategy that consumes grass as it appears, without regard to other site values, and relates solely to grass (not browse) and sites of low erodibility. These assumptions are often inapplicable (Table 3-2).

In practice, there are many situations where reducing soil erosion is a relevant goal, and several examples (such as in flood plains and less arid zones) where range management can be used to manipulate vegetation composition to favor productive species and reduce thicket-forming or poisonous species. In addition, in the whole of the semi-desert, 'opportunistic management' could kill the browse resource on which the economy depends. When these situations are added to those where mobility has been lost or where consideration must be given to wildlife or other values, the majority of projects will need to look beyond opportunistic management when setting environmental and social goals.

Table 3-2. Resilience of Rangeland to Opportunistic Management

Attribute of area under management	*Where opportunistic management theories apply*	*Where opportunistic management is risky or inappropriate*
Climatic zone	Arid to semi-arid	Very arid or semi-arid to humid
Landscape	Varied, with key resource sites	Uniform, lacking key resource sites
Erosion hazard	Low	High
Grass cover	Already depleted, low biodiversity	Diverse perennials, sensitive to grazing
Browse	Of secondary importance	Outvalues grass
Bush cover	Open stands	Thicket
Management objective	Stay within derived or deformed state	Move closer to natural state
Present mobility	Still mobile	Already confined

Note: Also see Annex C.

4

Enabling Policies

Successful pastoral development depends on ensuring a policy framework that recognizes a specific role for pastoralism. It is insufficient to rely on only a generalized rural development policy for several reasons — pastoral environments are stringent, forms of organization are distinct, and livestock kept by pastoralists have multiple roles.

Sectoral approaches flounder on the same obstacles. The most common approach has been to affiliate pastoralism with the livestock subsector, although sometimes pastoral areas are also regarded as contributing to the energy sector (as a source of fuelwood, or more rarely oil) and to the service sector (via wildlife and tourism). All this adds to the attention which pastoral areas receive, but in a disjointed manner and usually with marginal benefit to pastoralists or to NRM. Pastoralists are unlikely to benefit unless the guiding economic policy recognizes intrinsic value in pastoral economies and cultures, and unless it measures inputs and outputs in terms appropriate to those economies. The starting point is to recognize that pastoral value systems are not necessarily those of a monetarized agricultural economy.

Social policy, likewise, has to do more than label pastoralists as 'poor' or 'indigenous peoples'. Unless policies are formulated specifically to serve mobile pastoralists, education and health services will either bypass them or conspire to settle them, thereby accentuating social and environmental degradation. Much the same applies in the case of environmental and land policies — they too need to recognize the distinctive needs and circumstances of pastoral communities. Some of these points are covered in the next chapter as institutional issues. This chapter looks upstream, to the rationale for investment in pastoral development and to points of macroeconomic policy, including the case for subsidizing pastoral development.

Rationale for Investment in Pastoral Development

Investing in pastoralism presupposes balancing political, social, economic, and environmental considerations to arrive at the conclusion that investment is worthwhile. Occasionally there is clear-cut economic justification, but usually only in semi-arid or subhumid areas with a pastoral system that is already trade-oriented. More typically, a coincidence of other criteria that justifies intervention must be identified.

The strongest economic justification for intervening in arid zone pastoralism lies in the cost of *not* taking action — the imputed costs of structural famine relief or capital-intensive resettlement, compounded by increasing conflict and environmental damage. But this is also an example where justification lies in a coincidence of criteria. The fact that extensive pastoralism is more production per unit of land than alternative livestock enterprises is not, *in itself*, a justification for investment, since usually there will be parallel investment opportunities in other sectors that yield a higher return.

Political criteria often weigh against intervention, not necessarily by government intent, but more likely due to conflicting priorities or unrest in pastoral areas. In terms of other criteria, it would be a pity if such constraints were allowed to block investment. If other indications are favorable, it may be possible to avoid the problem by:

- proceeding with a study phase while a government resolves political constraints,
- making direct inputs to help government action, and/or
- adjusting the project area, either to exclude problem areas or to bring them under project influence.

Tactics would be determined by the nature of the political constraint. When the problem is insecurity, acute judgment is needed to fix limits for 'direct inputs', for example, supporting communities in self-protection but not supporting the suppression of unrest.

Social criteria provide the surest justification for intervention, especially in arid zones where hardship and deprivation are most evident. However, the justification is spurious unless based on standards appropriate to pastoralism. This means:

- avoiding bias towards mobile or settled lifestyles unless the ecological context (particularly the year-round availability of feed and water) is defined;
- judging degradation in relation to both context and cause (degradation caused by inappropriate water development, for example, may be reversible if agreement can be reached to use boreholes as drought relief instead of as the focus of slum settlements);
- relating poverty to stockwealth and the functioning of societal welfare support systems;
- recognizing where population pressure precludes any improvement in present living standards; and
- setting targets for standards of living accordingly.

Especially critical is recognition of intense population pressure. Although project intervention can proceed under these circumstances, the form of intervention must adapt to the severity of the problem and the options available for relocation. Defining the standards of living that are being sought could be made a matter of policy in project preparation.

Environmental criteria used to be applied regularly to justify intervention in pastoral areas, but that was before ecological rethinking popularized the notion that pastoral areas are not as overgrazed as commonly assumed. In reality, as discussed in the last chapter, most new projects will still have an environmental contribution, but it will vary according to the objectives of the project or project component. In particular, it is necessary to be clear whether the objective is:

- conservation, using NRM to shift the ecological status nearer to the natural state to benefit (a) national or other external interests, or (b) the pastoral economy (such as through wildlife management); or
- livestock support, using NRM within the derived state to benefit a pastoral system by (a) making best use of the grazing available, or (b) seeking additional benefit, for example, by preventing further soil erosion, improving vegetation composition, or focusing on a specific resource (such as browse).

As the objective varies, so also does the environmental justification for intervention.

The conclusion to be drawn is that as situations and criteria change, different types of intervention are favored. There are many variants, but four main types of rangeland intervention emerge that can be construed either as whole projects or as components in a broad-based intervention:

- Environmental conservation — focusing on an ecological attribute such as wildlife, without the immediate wishes of pastoralists being paramount.
- Livestock production — applying standard project procedures of the past on the grounds that there is sufficient commercial potential and investment content (and predictable outcome) to justify a straight investment project.
- Pastoral development — improving food security and the sustainability of NRM within the framework of an existing pastoral system using a process approach.

- Emergency operations — arising either as a form of disaster relief or in areas where the aim is to relocate pastoralists into other occupations.

Although it is the third category that is the focus of these guidelines, it is not the inevitable choice. All of the others have received World Bank support, and are not to be ruled out in the future. As the focus changes, however, so also does the nature of the project, so that the categories cited are not just variants on a theme (Table 4-1). The implications of this are examined later in the context of project portfolios (Chapter 12). The present focus remains on the third category, which has the widest application.

Economic Policy Instruments

Pastoral projects, like other projects, are much influenced by the macroeconomic environment in which they operate. This section provides an overview of four key points for review and perhaps amendment during project preparation. Their impact will be greatest in projects dealing with trade-oriented pastoral systems.

Exchange rate policies. Livestock prices are influenced greatly by the prevailing exchange rate in a country. Overvalued exchange rates lead to the import of cheap meat from the industrialized world, thereby depressing the income of pastoralists and increasing their vulnerability to drought. Depressed prices affect off-take rates.

The 1994 devaluation in the CFA[1] countries of francophone West Africa provides an example. Although devaluation disadvantaged urban consumers by substantially increasing the cost of imported goods, it helped local livestock producers by increasing the demand for local meat, thereby increasing offtake and livestock sales from pastoral areas to coastal markets (Box 4-1). The devaluation of the CFA did bring a 60 to 80 percent increase in the price of key veterinary drugs, and a 20 percent increase in the overall costs of production, but the effect of this was soon absorbed. In the Central African Republic (example in Box 6-1), veterinary drug sales slumped temporarily, but then recovered to reach $2 million two years after the devaluation.

Related trade policies. It is not only in West Africa that governments have allowed the import of subsidized ('dumped') meat from the European Union at prices significantly below local market prices as a means of providing cheap meat to urban populations. In clear cases of dumping, as covered under the definitions of the earlier General Agreement on Tariffs and Trade (GATT) and now the World Trade

1. Communauté Financière Africaine

Table 4-1. Features of Different Categories of Range-Based Projects

Feature	*Livestock*	*Pastoral*	*Emergency* [a]
Preparation	< 3 months	> 3 months	c. 1 month
Duration	c. 6 years	6-12 years	c. 2 years
Type	Blueprint	Process	Short-term
Funding	Standard	Flexible	Concessionary
Subsidy	< 20%	20-80%	> 80%
Justification	ERR 10-30% [b]	Social/NRM	Survival

Note: Excludes wildlife projects and (in the 'emergency' category) projects aimed at relocation of pastoralists.

a. Assuming a short-term intervention in disaster relief, *not* a long-term resettlement program

b. Economic rate of return.

Organization (WTO), the World Bank has included protection against such practices in its policy dialogue. Protection is best effected through the introduction of variable countervailing duties on the import of subsidized meat. For example, in Côte d'Ivoire countervailing duty was imposed on the basis of the price difference between imported and local meat of the same quality. However, if tarrifs are fixed too high at levels advocated in North Africa and the Middle East (200 to 300 percent), this can lead to heavy investment in livestock and purchased concentrate feeds, which in turn can greatly increase pressure on rangeland.

Crop input prices. Past pricing policies, especially in West Asia and North Africa (WANA), have favored crop production and disadvantaged rangeland-based pastoral systems. For example, tractors, agricultural equipment, and fuel were (and in many WANA countries still are) heavily subsidized. Over past decades, this has led to considerable crop encroachment into marginal areas where crop production would not be viable under normal market conditions. These trends have reduced the range areas in WANA over the period 1970 to 1988 by 13 percent (Glenn 1988).

Feed subsidies. The provision of subsidized feed became a feature of pastoral production in WANA during the 1970s and 1980s. It has led to dramatic increases in the total number of stock. For example in Jordan, which had a large feed subsidy program, the small ruminant population rose from about 1 million sheep in 1980 to 2.5

Box 4-1. Effects of Macroeconomic Policy on West African Livestock Trade

Reductions in the restitution payments granted on meat sold to West Africa by the European Union, reinforced by CFA devaluation, have restored competitiveness to local meat and boosted regional livestock trade. One year after these measures, regional livestock trade had overtaken non-African imports. Figures below illustrate this impact on European and Burkina Faso meat exports to coastal West Africa. The sustainability of this recovery will depend on continued improvement in production capacities and regional marketing efficiency

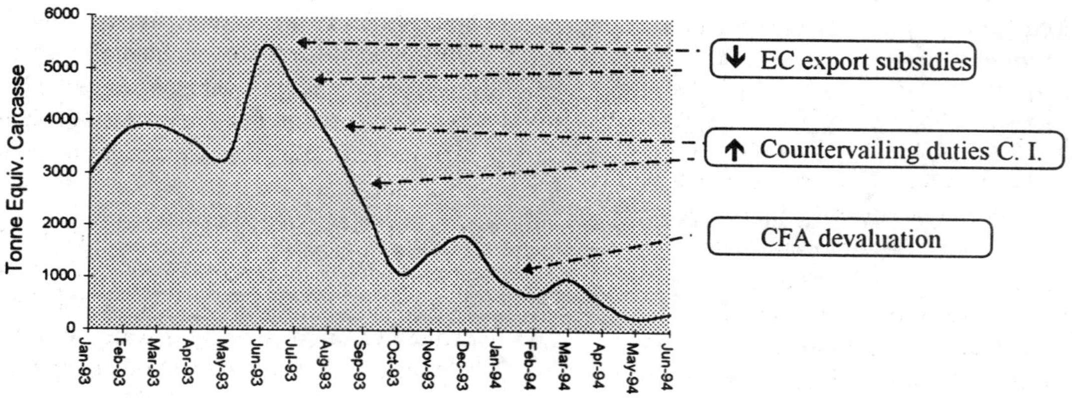

European Meat Exports to Ghana, Cote d'Ivoire, Togo, and Benin, 1/93–6/94

Compensatory increases in local (Burkinabé) cattle exports to the same countries were 59,563 head from January to September 1993, and 115,776 head during the same nine-month period in 1994.

Source: Vergriette and Rolland 1994.

million in 1993. At the same time, feed imports increased from 100,000 tons to over 600,000 tonnes. Under the influence of structural adjustment, most countries of the region have now abolished a regular subsidy, although subsidization of feed during times of drought is still common to the extent that it is becoming almost structural, effectively perpetuating the previous programs (see Figure 9-1).

The benefits of such subsidized feeding are doubtful. The policy maintains excessive stock on the range, which can be especially detrimental during the critical post-drought recuperation period. In line with 'opportunistic range management,' the policy should be not to *maintain* maximum stock numbers during times of drought, but to *destock* at those times. The combined effect of these measures in WANA is that the range livestock population increased over the period 1970 to 1988 by 28 percent. This has led to an increase in the average stocking rate from 2.8 hectares per Tropical Livestock Unit (TLU) in the early 1960s to 1.7 in 1988 (Glenn 1988). Careful scrutiny of all such policies is therefore needed before project interventions are designed.

Subsidy in Pastoral Development

Subsidy commonly features in the provision of government services to pastoralists. The case of subsidized feed just cited is just one example. The circumstances range from the redistribution of oil revenues to provision of famine relief. Usually, the underlying rationale is that pastoral systems lack the economic strength (or the economic orientation) to bear the full cost of services and development inputs, but social deprivation is also invoked, and sometimes environmental criteria. All these factors must be weighed when reviewing the justification for maintaining or withdrawing subsidies.

Economic justification for subsidy may be hard to find without placing a value on an environmental attribute. Direct economic justification arises in just a few cases, when:

- providing free protection against infectious disease (although then the benefits are national as well as local);
- causing pastoralists to maintain a breed more valuable to external economies than to their own (although ultimately market forces should prevail, without requiring continuing subsidy); or
- providing feed during times of drought to keep alive selected animals of particular value for rebuilding herds (although also with social justification).

Subsidizing feed supplies in non-drought situations is sometimes argued on economic criteria, although when this causes more animals to be kept than the available rangeland can support, it invariably causes environmental damage.

Environmental justification is invoked when an environmental attribute under pressure from pastoralists is deemed worth preserving. In practice, this is tenable only locally for a specific environmental attribute such as a wildlife habitat. Subsidy is wasteful if applied when the attribute of biotic or edaphic is depleted beyond recall, and potentially counter-productive if conceived as a blanket remedy to encourage pastoralists to keep fewer stock for reasons of general environmental conservation. (The reverse, however, of taxing pastoralists for animals in excess of subsistence requirements, may find a place, as noted in the next chapter.)

Social justification for subsidization is usually self-evident in the degree of social deprivation experienced by arid zone pastoralists. The most common area for subsidy is water development, although until recently most other government services were provided with little thought to cost-recovery. One point that will be elaborated in Chapter 10 is that field study is needed to judge the extent to which a society, and its various social strata, can be expected to contribute to cost-sharing. *As a general rule, subsidy should be concentrated into capital works and avoided in recurrent activities.*

5

Institutional Framework

To convert policies into practice requires budgetary and organizational provisions and an appropriate legal framework. The legal framework is considered first, focusing on issues of land tenure and the extent to which pastoralists are permitted to manage their own affairs. The advice offered here is general because a legal framework is very country-specific, and country statutes must be studied word for word to understand what is legally possible to help pastoral development (as shown by the case study in Lane 1996).

Land Tenure

Land law is usually written to codify and legitimize sociopolitical concepts of ownership and leasehold. Legislation follows to control land use, whether to avoid destruction of the land resource or to ensure that land use conforms to planning models. Where the State disallows or discourages the ownership of land, the second category of legislation dominates the statutes. Seldom is statutory law written specifically to legitimize customary law, although sometimes legislation is enacted to enable customary rights to receive legal recognition. Major differences can occur among neighboring countries (Box 5-1).

The value of land law in pastoral situations depends on the extent to which pastoral land is under threat from other users, and the extent to which those who aspire to take over pastoral land (including the State itself) are likely to abide by statutory law. Obviously, legislation will not control land grabbing if the parties concerned do not use or abide by the law. On the other hand, if there is nothing to stop the acquisition of land by powerful individuals or cliques, then there may be no point in proceeding with other lines of development. Familiarity with the content and effectiveness of land law is a necessary step in preparing for project intervention.

Corporate law and other legislation covering rights of association come into play when operationalizing participation. In an ideal world, it would be necessary only to delegate NRM responsibility to the appropriate decisionmaking body in a pastoral society, provide back-up services, and leave matters there. But seldom is that a realistic strategy.

In most situations where group responsibility involves holding money or land title, or taking decisions that may be disputed by persons not susceptible to societal control, the concerned group needs legal identity. This may be unnecessary if there are local government authorities competent to represent local group interests in court or in dealings with central government or commercial institutions, but seldom is local government geared to this task.

Two key questions about legal codes need priority attention:

- the forms of organization that are already recognized in law, and whether any of these have the simplicity and flexibility to be operated using existing customary procedures; and
- the possibilities for not just grass-roots organizations, but also for recognizing the authority of traditional leadership to keep organizations with responsibility for NRM within the jurisdiction of customary controls.

The case for reform is spelled out in most of the literature on pastoral development. The thrust is usually on comparing theories of tenure without reference to the many different types of pastoralism that occur in practice (Lane and Moorehead 1994). Within the total matrix of pastoral systems, there are situations where each of the current theories applies, even 'the tragedy of the commons'. Also, the case for reform is seldom expressed in terms attractive to politicians.

There is, however, a strong political argument in favor of land reform, since there is no doubt that:

- expectations *are* changing in pastoral societies,
- there *will* be increasing settlement,
- ad hoc settlement leads to impoverishment and discontent, and
- discontent is a threat to national security.

Hence action is needed if standards of living are to be acceptable in pastoral areas, and that action must include allocation of user rights since permanent water is highly localized and neither rainfall nor groundwater can be increased at will.

At that point, the argument flounders unless related to real localities. In principle, however, the argument would proceed to show how rights can be allocated, and how the situation might look after twenty to thirty years. One future scenario, for a dryland or semi-desert pastoral system that is now peripatetic, might be:

- residential areas designated where water is most abundant;
- preferential rights attached to outlying wells and their associated grazing areas;
- wet-season grazing areas shared among several groups;
- more of the population absorbed into paid employment;
- herding devolved to the minority (or to paid herders), following grazing patterns much the same as those now in use; and
- transferable grazing rights introduced to limit the number of animals using the range (especially those of absentee owners).

Box 5-1. Land Tenure in the Maghreb

The complexities of pastoral land tenure are starkly illustrated in the rangelands of the Maghreb (Algeria, Morocco, and Tunisia), where the divergence between statutory and customary law is compounded by very different state policies.

The present confused situation can be traced back to changes which occurred during and after the colonial period. Large areas were expropriated during the colonial period, and statutes governing collective ownership of tribal land and rights of cultivation or plantation were established. During the post-independence period, forests and rangelands were transferred to the State and private appropriation of collective land increased. Today, collective lands continue to be important in Morocco, whereas the policy of privatization in Tunisia has considerably reduced their importance. In Algeria, all collective lands are the property of the State.

Rangelands According to Land Tenures Regimes (percent)

Country	*Collective*	*Private*	*State owned*
Algeria	0	13	87
Morocco	59	←—— 41 ——→	
Tunisia	25	30	45

The functioning of traditional pastoral systems is now constrained by (a) conflicting pastoral laws, (b) numerous customary rangeland utilization practices, and (c) lack of coordination among the various actors. Future development in collective rangelands will have to support flexible positions which take into account all these issues.

Source: World Bank 1995a.

Whether or not events follow that particular development path is not the point, but it *is* important that:

- the pastoral society itself has a major say in determining the direction of change;
- government provides an enabling environment in a manner that encourages sustainable change;
- development planning addresses pastoralism in transition, not pastoralism as cocooned in a time capsule; and
- systems studies and process monitoring are allowed to guide the process (discussed in later chapters).

Changing legal codes must be identified as a specific task because too often perceived needs are expressed without reference to what legislative change they imply. There are two steps — first to specify the forms of tenure that are required, and then to make the appropriate textual changes to the existing legal code.

The proposed tenurial provisions need to be spelled out in terms of:

- the social groups and land areas to be given security (not easily defined below the tribal level);
- any seasonality of use and/or reciprocal arrangements that need to be accommodated;
- the form of security envisioned (freehold title or a form of leasehold or right of access); and
- the form(s) of organization that would hold rights of tenure, including the forms of constitution envisioned.

The review of the existing legal code must establish:

- where existing legislation suffices;
- which missing provisions require new legislation and which can be covered by instruments external to the law; and
- the draft bills required, including cross-checking through the legal code for inconsistencies.

This all takes time — probably at least two years from conceptualization to enactment — and is only worthwhile if the country has an independent judiciary to which the citizenry has access. Without such a judiciary, there is no basis for safeguarding pastoral rights and hence no basis for pastoral development.

The most critical step in the whole process of legal reform is probably when the ideas that emerge from participatory planning pass into the hands of those who draft laws. The latter are unlikely to comprehend fully all the nuances in the minds of the planners, and the planners seldom have any influence on the wording of the legislation once it becomes the property of the office in charge of legal affairs. The operational deficiencies of the Land (Group Representatives) Act, which was added to Laws of Kenya in 1968 to confirm the customary land rights of pastoralists, arose in the manner described.

Essential Services

Pastoral areas are usually among the worst served parts of any country. Where all-weather roads exist, it is more likely to be for reasons of national security or to serve tourism than to facilitate the flow of services and goods. Road improvement itself is not usually given high priority by pastoralists, however, and those who are mobile have a capacity that is often underrated to move essential goods (for example, diesel fuel for a borehole), provided that a supply point exists within the combined range of a truck and a camel. The most compelling reason to have roads in the view of pastoralists is usually access to medical services. It is basic social services that are given precedence below.

Mobile services are likely to need support prior to improvement in fixed facilities. The latter are needed in almost all districts, but should be organized within the framework of a regional plan. The tendency to site new facilities where some facility already exists must be resisted until growth centers have been selected. This is especially necessary in arid lands, where permanent water is sparsely distributed and limited in quantity. If a master plan does not already exist based on water supply and the other customary criteria of regional planning (demographic trends, land use, and axes of growth and trade), this must be organized at an early stage. Many district and local centers in the arid zone are already taxing finite water resources, and it is irresponsible to add to the load.

Health services usually separate human and animal health, even where it is manifestly illogical to do so. The two services must be kept separate down to the district level, but primary health care at lower levels could often be improved by unifying delivery.

Literacy requires early attention because, as participatory planning and development progress, more responsibility passes to those in the community who are literate, and those not literate are liable to be sidelined. The World Bank's project record already shows increasing attention to literacy so there is no need to dwell on this point, but it is important to initiate adult literacy programs at the first opportunity and focus these and other training programs on issues and tasks with which the community will engage (Box 5-2).

Education has dimensions far beyond literacy. Although project involvement may be confined to adult literacy and the skills needed for communities to participate in development, there is scope for much more, both among pastoralists and those with responsibility for assisting pastoral development. The rationale was outlined in Chapter 1, and further specific suggestions appear in Chapter 12.

Technical services include extension, the back-up needed in research and ensuring animal health and water supplies, and provision for marketing, banking, and credit. All these require attention during project preparation. An early priority would be training and retraining in order to produce people competent in pastoral extension and the provision of back-up services. Furthermore, the need for credit must be assessed as a policy issue because expensive items such as boreholes may warrant subsidy.

Institutional Arrangements for Development Support

Much hinges on the institutional arrangements for the actual delivery of services and inputs. These arrangements need scrutiny as close as any part of the development process. A wide range of services is involved, from counseling on how to make use of available legislation to ensuring that needed drugs are available.

The available options can be summarized in terms of five models:

The departmental delivery model delegates to each line ministry responsibility for the delivery of its part of the program. The model may assume that local authorities oversee or coordinate the program and that the program

Box 5-2. Experience with Literacy in World Bank Projects

In Senegal, functional literacy was seen as an important requirement to achieve producer participation in the management of herder associations. From the start of World Bank assistance in eastern Senegal in 1979 to its end in 1989, some 9,000 people passed through the literacy program (over 10 percent of the population). The literacy campaign proved effective to train livestock auxiliaries capable of keeping accounts for the herder associations for their purchases of drugs and cotton seed. The presence of livestock auxiliaries and their pharmacies was regarded by local herders as the major contribution of the projects. There is little information, however, on who benefited from the literacy campaigns (in terms of age, gender, and socioeconomic category). In Senegal, literacy training and materials were mainly in Pulaar, the vernacular language of the pastoral peoples.

In the Central African Republic, the Fédération National des Eleveurs Centrafricains (FNEC) continues to provide adult literacy training for Groupement d'Intérêt Pastoral (GIP) members, especially women, and train them to run the GIPs. After a period during which the choice of the language was an issue, this training is finally conducted both in the Foufoulde and Sango languages.

Source: World Bank internal documents

originates from proposals from the grassroots level, but ultimately what happens (or not) is determined by the will and efficiency of individual ministries. This is often regarded as the best model to develop technical skills and to ensure post-project sustainability, but in practice it perpetuates compartmentalized and top-down thinking, often leading to misconceived and poorly coordinated development support.

The coordinating department model designates one department with responsibility to coordinate and ensure delivery of inputs from all relevant sources. This offers the possibility of a coordinated flow of inputs, but the relevance of the flow is still suspect and its delivery is only as good as the budget and the influence of the coordinating department. A department close to the political center would have influence, but may still have difficulty ensuring that technical ministries define relevant inputs.

The unified area authority model establishes an extra-ministerial authority with overall responsibility for organizing all inputs in a designated geographical area. This is often the most efficient model to assemble a motivated staff and meet targets. But it is commonly associated with an autocratic style, more proactive than demand-driven, and often attracts only grudging support from the rest of the government machine. It is often not sustained once external support has stopped.

The local government model attaches the role of a unified area authority to local government so that services are brought under the control of the democratically elected representatives of the people served. This is an unpopular notion with central governments because it implies relinquishing power from the center, although local government structures are usually in place and looking for a more meaningful role in development.

The service cooperative model involves herder groups directly in delivery of private goods (such as animal health and water services), and eventually in sub-contracts from the government for public goods (such as education). The idea of producers forming a service organization is tried and tested, mostly in higher potential or more commercialized situations, but also successfully in Mauritania and Niger. It is important, however, to keep the cash-flow of commercial and social functions separate. Commercial functions should be self-financing, whereas social and public services should be funded through general levies or transfers from the treasury to the service organization. Disappointing experience with service organizations almost always stem from mixing social and commercial objectives.

Decentralization is a means of devolving responsibility, facilitating democratic processes, applying principles of subsidiarity, and encouraging participation.

The whole process of devolving responsibility to communities and local organizations is considered here as 'decentralization' since most government systems start as being heavily centralized. There are two avenues to consider — allocating responsibilities downward to local government and beyond, and building the capability of communities to reach upward to contribute more to the development process.

Local government was portrayed earlier as having a stronger role to play in pastoral development than is commonly allowed. This will not readily be achieved because the present balance of power and responsibility is orchestrated by the central government. A shift in position will only occur if the higher echelons of government perceive it to be in the interests of good government and able to be put to political advantage. If so, the following initiatives could be among the most practical ways of strengthening the role of local government:

- District planning should take precedence over centralized planning. This procedure is already adopted in some countries, requiring ministries to base their sectoral plans on integrated plans from the district level. As long as district plans remain the product of district-level ministerial staff, however, it is difficult to break the sectoral mold and obtain a real expression of community priorities.
- Staff secondment from central to local government should be considered as a

strategy to improve the competence of local authorities. After pilot testing, this could be considered as standard procedure. It should help integrated district planning and also provide wider benefits by bringing technical staff into closer association with the people they are expected to serve.

- Training of councilors and others associated with local government (including herders themselves, so that they can participate as equals in the planning process) should be stepped up so that these officials and representatives are better equipped to contribute to pastoral development.
- Local government revenue may be an area requiring external advice, examining revenue collection and reducing central subvention. One possible source of revenue is the introduction of a property tax on livestock, applicable where the individual or family holding exceeds a level calculated in relation to subsistence requirements. This could specifically target absentee owners. Undoubtedly the response would be to transfer animals to the ownership of poor families, but this may not be bad if it serves to provide those families with better food security.

Charities and other NGOs have repeatedly shown their ability to assist at a community level to facilitate development. They also provide a means of channeling funds and goodwill from those with a social conscience to those in need. No doubt these organizations have much to contribute to future pastoral development. However, they also have limitations. They are often unrepresentative, absorbing large inputs for only localized benefit and with only sketchy evaluation. Care is needed, therefore, not to place so much reliance on NGOs that efforts are relaxed to improve the support capability of democratically elected local bodies, both local government and community-based organizations.

Community-based organizations are considered later in the context of herder organizations (Chapter 7). The attraction to central government of sponsoring their establishment will lie largely in their ability to facilitate cost-sharing. The main contribution that central government can make to their establishment is to extend formal recognition to customary organizations, and ensure that simple forms of organization are available to meet the functions outlined in Chapter 7.

The commercial private sector should also be encouraged to participate where feasible in the delivery of services to pastoral areas. The sorts of services that might be provided in this way are considered in more detail in Chapter 8. The first step in that direction is to establish a clear understanding of the distinction between public and private goods (Table 5-1).

Table 5-1. Institutional Framework for Natural Resource Management

Service or good	Type of good or service	Responsibility: State (national) level	Responsibility: Local government	Responsibility: Private associations	Individual herders, commercial and private entrepreneurs
Basic infrastructure, including basic physical infrastructure and social services, as conditioned by regional conditions					
Physical planning (at regional level)	Public good	Regional planning framework based on resource distribution and use, demographic and economic trends, and water availability.	Surveys to update and improve regional plans, issue of zonation and other permits according to agreed plans.	Participatory input into planning.	Studies contracted out to consultants.
Primary road network	Public good	Trunk roads serving all main administrative centers.	Shared responsibility for trunk and toll roads, organize in (cooperation with others) food-for-work road building.	Link roads serving individual association markets plus cost-sharing for access to remote areas.	Subcontracts of road construction and maintenance to private companies.
Other communications	Toll goods	A policy framework that facilitates private inputs and regulates public inputs into the expansion and modernization of communication.	n.a.	Strong participation in the design and content of radio messages.	Inputs by the commercial sector in the expansion of the communication network.
Basic education	Public good, some specialized education is private good.	Teacher training and primary education, based on targets and curricula set regionally.	Basic education adapted to the needs of the pastoralists, including adult literacy.	Inputs, often substantial, from religious organizations and other NGOs.	Koranic schools and other private training for which beneficiaries choose to pay.
Potable water	Public good (where natural sources are not already in place).	Policies and provisions that set regional targets for water quality and supply.	Supervision of the implementation of regional water supply targets	Collective planning and operation of water points by associations.	New facilities installed by contractors under cost-sharing arrangements, and private wells maintained and improved.

Table 5-1 (continued)

		Responsibility			
Service or good	*Type of good or service*	*State (national) level*	*Local government*	*Private associations*	*Individual herders, commercial and private entrepreneurs*
Primary health care	Public good with externalities for most preventive actions, private goods for most clinical treatment of individuals.	Policies and procedures that ensure adequate protection against epidemics; enabling environment and quality control for the sale of pharmaceuticals.	Regional supervision of the quality of disease prevention, basic education in hygiene, child and mother care, etc.	Distribution of pharmaceuticals and further health care programs.	Develop, produce, and distribute pharmaceuticals; plus basic animal health care systems provided by private operators.
Specialist services, including inputs most commonly associated with economic development in the range livestock sector					
Specialized education	Mostly private but with strong enabling role from public sector.	Establishment of centers of excellence on the basis of population density and stage of economic development.	Establishment of appropriate (mobile) centers and vocational training.	Management of facilities.	
Water development	Public good with externalities	Establishment and supervision of appropriate water policy.	Control of the geographic distribution of water points, monitoring of water resources and abstraction.	Main responsibility for the operation and maintenance of group water supplies.	Responsible for construction of private water points, operation of private wells; subcontracts for communal/group water supplies.
Preventive animal health services	Public good with externalities and moral hazard issues	National disease control strategies, border protection and overall responsibility for disease surveillance, quality control on vaccines and veterinary pharmaceuticals.	Monitoring of disease outbreaks; supervision of compulsory vaccination campaigns, regional quarantine measures, inspection of livestock products.	Participate in policy and implementation, including compulsory vaccination campaigns.	Produce vaccine and pharmaceuticals by commercial sector, subcontract most preventive animal health tasks to private professional or paraprofessional veterinary agents.

Table 5-1 (continued)

		Responsibility			
Service or good	*Type of good or service*	*State (national) level*	*Local government*	*Private associations*	*Individual herders, commercial and private entrepreneurs.*
Curative animal health interventions	Mostly private goods	Limited to overall supervision (plus some inputs to training).	Training and monitoring of private paraprofessionals.	At the regional level, responsible for drug distribution; at the grass roots level, responsible for servicing/supervising paraveterinary agents.	Combinations of private sector veterinarians and paravets in herder associations can be envisioned.
Animal production services	Mostly public goods with externalities (private goods increasing with commercialization).	Conservation of genetic resources in cooperation with breeder associations, development of appropriate advisory policies and systems in livestock production and NRM.	Provision and supervision of marketing structures, shared responsibility with associations for advisory services.	Operation of marketing infrastructure, supervision and monitoring of extension.	All other production and marketing operations.
Research	Mostly public good, private where developer can capture benefits of the research products.	Establishment of priority-setting mechanisms to ensure a balanced program, reflecting zonal needs and opportunities.	Ensuring participation by stakeholders in priority-setting, supporting relevant research on arid range-livestock production.	Major role in priority-setting and eventual limited degree of cost-sharing.	Inputs only with high value products where it is worthwhile for the private sector to invest in research (e.g., vaccine and drug development).
Drought monitoring, mitigation and recovery	Public goods at regional level; private goods at on-farm level.	Establishment of national drought preparedness policy and network, follow-up actions through national food banks and support for accelerated off-take in case of widespread drought.	Supervision of data collection and local analysis, responsible for food aid distribution and supervision of restocking programs.	Data collection in cooperation with local government, use of traditional support mechanisms in case of droughts, participation in credit and insurance schemes for destocking and restocking programs.	Inputs to handle the accelerated off-take of livestock in times of drought and provide animals for restocking.
Other services	Mostly public	Generation of alternative employment opportunities	Monitoring wildlife resources.	Some scope for joint ventures in ecotourism and other non-livestock resources of arid lands.	

Table 5-1 (continued)

		Responsibility			
Service or good	*Type of good or service*	*State (national) level*	*Local government*	*Private associations*	*Individual herders, commercial and private entrepreneurs.*
Law and order are usually taken for granted in development planning, but often call for specific attention in pastoral areas.					
Allocating and ensuring land rights	Public good with externalities	Establishment of national land tenure policy and legal framework.	Survey and registration of customary land rights.	Operation and adaptation of customary tenure systems within the framework of the law.	Subcontract surveys to consultants.
Maintaining security	Public good (unless insecurity an internal issue, of little/no national significance).	Maintaining and deploying national security force, including assurance of access.	Monitoring security situation and alerting national security forces in case of incursions.	Customary control of incursion; possibly by delegated responsibility, border control.	n.a.
Preventing criminality	Public good with externalities	Establishing and upholding criminal code.	Enforcement of criminal code in parallel with customary law.	Imposition of customary codes of sanctions to the extent that they do not conflict with the civil code.	Parents (and Koranic schools, etc) instill codes of ethics in the young.
Promoting the civil code	Public good	Establishment of civil codes that encompass subsidiarity and reflect societal ethics.	Enforcement of civil code in parallel with customary law.	Maintenance of customary ethics, use civil courts in disputes (usually fiscal or territorial, invoking statutary law).	Parents (and Koranic schools, etc.) instill codes of ethics in the young.
Promoting subsidiarity	Public good	National policy, allow self-determination through democratically elected local bodies; register responsible bodies.	Assumption of delegated responsibilities; review processes.	Maintenance of customary procedures (pastoral societies are well versed in subsidiarity); fulfill new responsibilities as delegated by the State.	Herding families exercise customary and delegated procedures.
Promoting associations	Public good	National policy and legal framework, including provision of adequate independence of pastoral associations	Catalyze and support creation of new associations and act as an intermediary between associations and national government.	Maintenance of societal hierarchies and groupings effective in the areas of decisionmaking for which they are designed; new forms of association as appropriate for new activities.	Contributions as members and officers of associations.

6

Essentials in Project Design

All projects need information on which to base their planning, as well as participation in order to ensure relevance and local support. And to ensure sustainability, the packaging of a project (in terms of timing, flexibility, etc.) is no less important. These needs vary, however, according to the orientation of the project — the relative emphasis given to economic, environmental, and social objectives, and the scale of operations.

The implications of scale are dealt with as they arise, but the orientation is assumed to be constant, and directed at:

> *enabling pastoralists to adapt to their changing environment in ways that improve food security and the economic choices open to them, while sustaining the capacity of the natural resource base to support pastoralism or future forms of range resource use.*

This objective implies an open-ended development process. Projects supporting the process would not be open-ended, but they would ensure the instruments of change for an initial period until institutions are in place to enable pastoralists to influence future events. Such projects need a monitoring system to adjust inputs based on experience. Monitoring implies that data gathering will continue throughout the project term. There would still be a need for baseline data to guide and justify the initial inputs, but less reason to know everything at the outset concerning the pastoral system(s) involved.

Baseline Information

The information required to start a project falls into three categories: (a) the database needed to prescribe a project area, (b) the information to collect during field appraisal, and (c) topics likely to require further study.

Project area needs careful definition because pastoral development can be tackled through projects of very different scales, including:

- national projects that provide the overall institutional support needed for government and community action;
- district-level projects that address the development needs of a specific administrative area, usually through a mixture of infrastructure and field-level inputs; and
- community-based projects that focus on one or more pastoral groups and address wider issues only to the extent needed to facilitate progress in the group area.

Sometimes these scales get mixed within one project — often with a conflict of interests (as in the Central African Republic, Box 6-1) — but the basic divisions are well known. Sometimes pastoralism also receives attention indirectly through sectoral projects (such as education or tourism) or under gender or poverty initiatives. In all situations, however, a correct definition of participants and project area is essential.

A project assisting pastoralists in NRM must embrace the entire area used by a pastoral population, including all of their seasonal grazing areas and drought reserves. The only situation where it is straightforward to fix a boundary for project intervention is where the focus of attention is a semi-sedentary pastoral system set in an area of relatively reliable rainfall. In other situations, the outer limits of project intervention will not be clear until *after* the first round of participatory discussion.

Even if there were a mass of information on ecology and pastoral practice, it would be unsafe to define a perimeter until there has been consultation on customary mobility and drought strategies. This applies even where the project area is nominally prescribed as an administrative district.

Data collection is more time-consuming in pastoral situations than in most fields of rural development. Much of the required information comes from participatory rural appraisal (PRA) and ensuing enquiries, as discussed below in the context of participation. At the outset of project identification, however, the net must be cast wide because then there is no 'target' population and the field itself is uncomfortably wide with legal aspects, environmental sensitivity, and special needs in water engineering and veterinary medicine.

Sources of information need to include non-government as well as government institutions. This implies tapping universities, remote sensing

Box 6-1. Delivery of Livestock Service in the Central African Republic

The National Federation of Central African Livestock Producers (FNEC) was formed by herders in the 1970s to fill the gap left by the collapse of the public sector in the Central African Republic. It was rehabilitated and took effective responsibility for veterinary drug marketing and distribution in 1982 (under the World Bank assisted Livestock Development Project). Concurrently, training programs were reoriented toward teaching herders the proper use of these drugs, and members of the FNEC were trained as auxiliaries to supplement the clinical and preventive services provided by the public sector.

Total sales of veterinary drugs and feed increased dramatically after being taken over by FNEC. By the end of 1986, sales had jumped to US$ 1.7 million (from $9,000 in 1981), and to $2 million at the end of 1992. FNEC's general success in drug imports and distribution activities encouraged individual herder associations to expand into the sale of other inputs. In 1989, FNEC began providing extension services and expanded education programs for its members (under the National Livestock Project).

Lapses in management, combined with political interference and price controls, led to a decline in effectiveness in 1992. However, FNEC's statutes were revised in 1993 to ensure majority grassroots representation, widen its sphere of activity, and confirm its non-political character under the recently launched Livestock Development and Rangeland Management Project (LDRMP). In 1995, sale of veterinary drugs recovered to reach 1992 the value again.

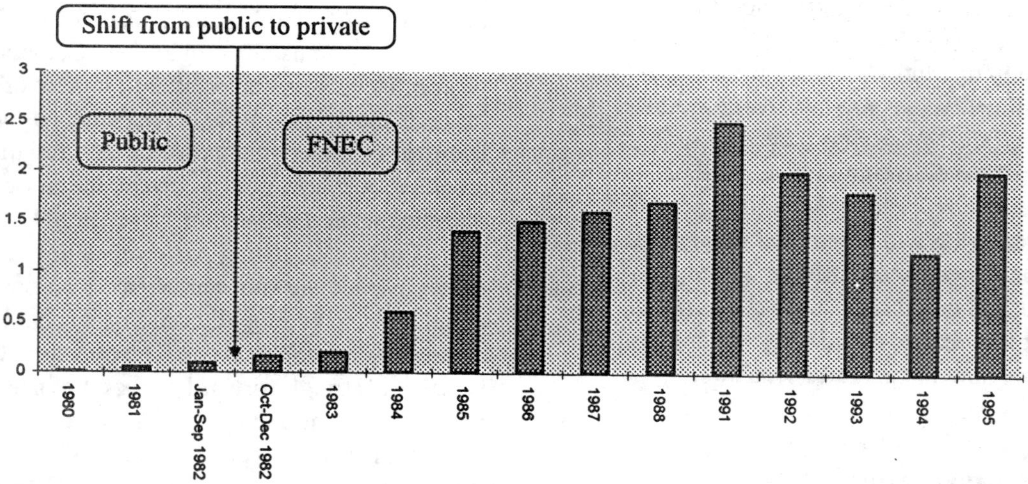

Value of Veterinary Drug Sales (and Other Inputs) in the Central African Republic, 1980–1995 (sales in millions of U.S. dollars; data for 1989 and 1990 are missing)

Source: Umali, Feder, and de Haan 1992; personal communication 1993-1995

units, and donor agencies (not least NGO offices), as well as informed people associated with existing pastoral networks. Pastoral networks operate both internationally and in several countries. They are active in information exchange and often as lobby groups. While the representation of pastoralists in such forums is limited, consultation with network members provides insights on current pastoral organization and issues. Such sources can also advise on the full spectrum of government institutions that need to be consulted. Sometimes there is one coordinating unit, but otherwise obtaining a government perspective on pastoral development involves a lengthy round of consultation. Because there is often a rapid turnover of staff, it can also be difficult to obtain a full spectrum of background data from government sources.

Steps in data collection pass through three levels of study. The first level uses the existing database and PRA to define whether the situation under consideration requires:

- an emergency operation or abandonment of plans,
- a standard project with clear-cut investment goals, or
- a process project of the type needed in most pastoral situations.

Assuming a process approach is selected, additional data collection (the second level of study) can fit into the project as it proceeds. Process projects are described in more detail later in this chapter, but in the present context are characterized by having a built-in monitoring system to ensure that the project receives the flow of information needed to keep it moving in an appropriate direction. From a financial perspective, the process approach is a procedure that allows investment in research to be deferred until it can focus on relevant issues and inputs, at the same time avoiding wasteful expenditure on inputs that otherwise may be made unnecessarily, prematurely, or detrimentally. Were it not for the process approach, a much heavier investment in research would be needed prior to the project.

The third level of study — systems studies of sufficient scope and detail to analyze the responses of pastoral systems to their fluctuating environment — is usually beyond the scope of project preparation. Process monitoring is a good substitute if continued long enough, but opportunities should be sought in parallel to ongoing development projects to encourage additional systems studies. This topic is pursued in Chapter 12.

Key data fall into three categories:

Environmental data. Planning depends on being able to relate territories to ecoclimatic zones. If there are no ready-made zonal maps, then original thematic data need to be scrutinized:

- climatic data, particularly long-term rainfall records, need to be examined for seasonality and periodicity;
- geomorphological maps help by differentiating landscape types (such as basement, laval, and sedimentary surfaces);
- hydrogeological data are essential to define the potential for expanding settlements and dry-season grazing areas;
- satellite imagery can be useful in the field to stimulate discussion with herders on resource use and production seasonality; and
- vegetation maps form a useful reference in PRA discussions and in delineating key resource areas (although their use in predicting carrying capacities and development strategies is usually misleading).

Resource use statistics. Most statistics are recorded by administrative area but usually can be broken down to give an idea of the status of rangelands in and around the prospective project area. Most relevant are:

- territorial limits of groups as recognized by government;
- legal status of land (State, trust, and allocated land);
- the social strata requiring attention, with particular reference to gender, caste, and wealth;
- recent land use changes (for example, crop encroachment in key resource areas);

- human and livestock populations (including, if possible, trends broken down by age and gender); and
- input/output data (such as livestock sales, seasonality of price fluctuations, cereal/livestock price ratios, famine relief, etc.).

Categorization of pastoral systems. While it is essential to know which pastoral systems operate in and around the proposed project area, it is rare to find a categorization in use along the lines tabulated in Chapter 2 — information must therefore be gleaned and put together through PRA and subsequent studies.

Participation

Participatory planning involves at least two stages — the initial rounds of enquiry and discussion, usually through participatory rural appraisal (PRA), and the more detailed studies that are needed to decide specific inputs. The normal way of arranging participation is to:

- hold discussions with the local leadership and any existing associations,
- hold open meetings according to customary procedures,
- form small groups of individuals with interest and acknowledged expertise for initial detailed discussions,
- use this experience to identify (a) interest-groups not initially covered and (b) individuals for in-depth review of specific issues, and
- take ideas and conclusions back for community review.

Participatory rural appraisal (PRA) is now a well-documented and well-tested procedure (Niamir 1990; Waters-Bayer and Bayer 1994). A summary of concepts and methods is included in Annex E, and an example of recent World Bank experience is added in Box 6-2. There are, however, two problems in PRA that need to be stressed.

Survey fatigue is evident in many communities which have been subject to repeated social enquiry for little or no return. In such cases, it is only the most vocal members of the community and those with vested interests who volunteer to participate in PRA. To avoid this trap it is necessary either to start with such high-powered publicity that everyone believes participation will be worthwhile, or to begin quietly and slowly and diffidently in order to build personal rapport. The latter is probably the better strategy, but to be effective, it requires time, personality, and linguistic ability on the part of those who are initiating the process.

Superficiality, although generally recognized as inherent in PRA, is a particular problem in pastoral situations. Apart from sensitivities in specifying livestock numbers, there are many pastoral practices, including mobility, so deeply embedded in societal and environmental lore that it takes a long time to reveal their rationale. Nor is it easy to separate the perspectives of overlapping pastoral groups, or sometimes social strata within one group, without subjecting these to separate enquiry.

Pastoral PRA is likely to comprise several steps and lead to follow-up studies. The lines of enquiry to be pursued during PRA are indicated by the list below, which itemizes topics commonly requiring detailed study. The output of PRA, however, should be more than an agenda of further studies. PRA should establish whether there is a basis for proceeding with a project, and if so, whether there is an investment component that can precede further studies.

Detailed studies can be expected in three principal areas. Some, such as modeling, are iterative and open-ended, but most should be completed in a year.

Modeling of the pastoral system(s) helps to clarify links between the resource base, climatic rhythms, NRM practices, and family and societal goals. A first approximation would arise from PRA, but invariably the initial model can be expanded and improved. It can be useful to incorporate data into a mathematical model, although the models developed within a participatory framework would be expressed exclusively in words and pictures. The inputs to modeling would come from the individual studies indicated below.

Building blocks of the pastoral system(s) need to be firmly established, with particular reference to (a) diet, including adaptations to drought; (b) the basic structure of social-territorial organization and resource tenure; (c) conflict resolution, decisionmaking, and sanctions; (d) welfare support mechanisms; and (e) economic links with market economies and other groups.

Components of NRM requiring detailed study include:

- imposed and voluntary components of movement (Annex A),
- uses made of key resource areas,
- complementarity of the livestock species present,
- non-livestock uses of vegetation,
- drought strategies, and
- relationships between NRM practices and stockwealth.

Participatory development must be the outcome of participatory planning. It is the subject of later chapters, dealing both with forms of local

Box 6-2. PRA in the Matruh Resource Management Project (Egypt)

The Matruh Governorate is located in Egypt's western desert. The majority of the inhabitants (85 percent of 250,000 people) are Bedouins who still maintain a traditional society, but one that has been modified during the last decade as people adopted a more sedentary lifestyle. When a Bank-assisted project came under consideration in 1990, World Bank staff argued that a traditional livestock project was not appropriate, and recommended that central and local government officials collaborate with the Bedouins in identifying and preparing the project.

A local task force with seven teams identified the project using PRA techniques applied over a three-month period. Meetings with women were held by a female consultant teamed with a female Bedouin veterinarian to cover the same ground covered by the all-male meetings.

The local task force and the local community took about one year to prepare the project. Consultants and World Bank staff were present only intermittently. The resulting project contained the following components:

- *natural resource management* aimed primarily at conserving the water, soil, and vegetation of the Matruh Governorate;
- *adaptive research and extension,* focusing on dryland farming and livestock production systems, range management, sustainable agriculture, and training directed at the local communities;
- *rural finance* for small farmers, the landless, and rural women to help them engage in income-earning activities.

Community action plans are being prepared by community groups built upon traditional Bedouin lineage structures. These community groups will be involved in implementing the plan and monitoring the results.

Participatory planning enabled: (a) community empowerment and action, (b) trust and respect between Bedouin and Government experts, (c) ease of negotiations (just over 3 years elapsed from the first identification mission to effectiveness), and (d) community commitment and the prospect of sustainability.

A weakness is that the project description in the loan documents is still too prescriptive and lacks the flexibility to allow for changing perspectives and innovation by individual communities. Similarly, monitoring indicators still focus on physical achievements and not on progress achieved in beneficiary involvement. Thus, more work is required to ensure that participatory planning translates into participatory development, with flexible project documentation and the monitoring of well-identified and relevant process indicators.

Source: World Bank 1995d

organization and individual project components, and so warrants only passing reference here. The point to be made here is that preparations for the establishment of appropriate forms of organization must start at the earliest possible stage, by:

- introducing communities (especially their leaders) to the options open to them under existing land and corporate law,
- discussing modes of representation and the types of group constitution that would suit their circumstances, and
- possibly inviting them to try, in an informal manner, the forms of association which are expected to be introduced formally later.

How long it takes to proceed from PRA to investment depends on circumstances, but three to four months of initial PRA and an additional six months following up on key issues, should form a reasonable basis for initiating an investment project. It is, of course, easier to make an early start if the initial investment component focuses on inputs that are ecologically and socially 'safe', and yet help toward more sustainable NRM. Some ideas are given in Chapter 10.

Sustainability

In the context just cited, 'sustainable NRM' refers principally to environmental sustainability. Projects are also expected to be sustainable, however, in the sense that their initiatives are maintained by governments and communities. But in either case, the notion of sustainability that is imputed is not absolute but related to our capacity to predict change. A sustainable pastoral system might be expected to cope with a few years of drought but not with a decade without rain that might result from a climatic shift. Methods for achieving relative sustainability are considered here.

Physical sustainability (ensuring that project initiatives last) is sought in all projects by scaling inputs at a level deemed to be manageable under post-project conditions. However, two ploys have particular significance in pastoral development projects.

Building community participation is especially important. Central governments usually have great difficulty servicing pastoral areas, so the more that communities can do unaided, the more sustainable are project initiatives.

Building local government capability works in the same way by enabling locally elected bodies to provide services that central governments have difficulty delivering. This point is elaborated in Chapter 8.

Environmental sustainability is an inherent objective when intervening in NRM. The chances of achieving that objective decrease, however, as rainfall reliability decreases, population pressure increases, and freedom of movement diminishes. Also, the feasibility changes according to the goals that are set — whether to stay somewhere within the derived ecological state or to move closer to the natural state. A statement of these goals and the standards of living being sought is basic to the design of a project that seeks environmental sustainability.

Generalizing, there are three situations that a project may face:

(1) Where population pressure is not compatible with any notion of environmental sustainability, and where progress requires:

- an increase in land area (usually not feasible) or other outlets to support part of the population, or
- routine famine relief (perhaps eased by project inputs).

(2) Where there is some scope for supporting and supplementing customary NRM, but where the resource base is eroding with little chance of arresting the decline without:

- recourse to one of the above strategies, or
- heavy investment in land rehabilitation.

(3) Where there is no major obstacle to development, but where different targets or strategies are required depending on whether:

- resources are valued solely for pastoralism, or
- wildlife or some other value is added.

One factor that will influence how easily goals are attained is the provision that is included to accommodate dry years and drought. A number of drought management strategies are outlined in Chapter 9. Even so, rainfall remains unpredictable, and there are few goals that can be realized through a blueprint for action that does not adapt to circumstances. A monitoring system is an essential complement to any environmental strategy.

It should be noted that the situation where resources are valued solely for pastoralism is taken as the norm for the specific guidelines that follow in Part Two, with occasional reference made to other situations.

Process Approach

Projects that are intended to assist development processes must have a different style from those whose job it is to implement a pre-formed blueprint. How a process approach differs from a blueprint model is shown in Table 6-1.

Project identification is based more on an overall assessment of development goals and priorities than on isolating sectoral opportunities and quantifying potential inputs and outputs. Need may be given more weight than opportunity. The product of identification is:

- a relatively comprehensive although mainly qualitative statement of the present situation expressed in environmental, social, regional, and economic terms;
- a representation of likely development paths, indicating assumptions and incremental inputs at each stage; and
- costs and justification for the initial inputs proposed.

Charting likely development paths is essential and a task in its own right. There is nothing unusual in thinking through the direction that development is taking, but it is not so common to do this formally as a means of deciding sequences of incremental inputs. Most pastoral situations require not one development path, but several in parallel for:

- different social strata with variable systems of livestock production and NRM;
- different areas that vary in their potential and season of use;
- individual elements of the system, charting the order in which constraints to livestock production or a major institutional issue would be tackled; and
- two or three alternative assumptions about demographic change or land or fiscal policy.

Nothing very complicated is required. The aim is to state where a system (or component) starts, where it is going — particularly demographically and in terms of lifestyle — and what needs to be done to keep moving in the

Table 6-1. Comparison Between Process and Blueprint Approaches

Blueprint approach	*Process approach*
Project identification	
Focus on sectoral constraints and opportunities	Focus on constraints and opportunities as they affect farmers or pastoralists, leading to a statement of the present situation and likely development path(s)
Project appraisal	
Focus on physical inputs (boreholes, etc.) and outputs (meat, etc.)	Focus on first phase inputs and management of processes and risk (after confirmation of goals and validation of proposed actions)
Project implementation	
Assess progress against the appraisal report through prescriptive monitoring	Focus on adapting inputs (and development paths as necessary) on an annual basis, guided by process monitoring

desired direction. Institutional as well as technical inputs (including research) are scheduled with consideration for the time needed to get results.

Preparation of development paths must be a participatory exercise. Planners will have their own ideas on what is important — for example, the flow of benefits to poor families and women and children — but it is important that the priorities of the community are given full weight. One of the main advantages of the method is that it helps avoid ill-considered steps by forcing consideration of where each step leads, both immediately and in the longer term under different scenarios. Participatory preparation is educational for both sides.

Project appraisal is undertaken on the basis of the presentation indicated, supported by further consultation (and PRA) as necessary. Particular attention would be given to:

- assessing the comprehensiveness of the situation report and the validity of proposed phase one inputs, including their projected economic, environmental, and social impacts;
- confirming or elaborating environmental goals and projected impact on standards of living;
- ensuring arrangements for implementation, especially to establish community-based organizations, support systems, and an appropriate monitoring system; and
- recommending a budget for phase one, presented within an indicative framework for a five- to ten-year period.

Implementation proceeds on the basis of the appraisal report, with provision for annual (or at most eighteen-month) reviews of progress, at which recent monitoring data and participatory consultation would establish the forward program and budget for the next twelve to eighteen months. There is *no* blueprint, and the development path is used (and revised as necessary) only as a guide to make forward commitments. Supervision missions would review progress and conceptualize adaptations to the plan against actual monitoring data and not just the appraisal report.

Part Two
Guidelines for Specific Project Components

The next five chapters focus on individual project components. Chapter 7 concerns herder organizations. These are the organizations — both customary and introduced — through which herders organize and receive development inputs and exercise NRM. The chapter covers a range of organizational forms, and assesses their suitability for particular situations and services. Chapter 8 concerns mechanisms to support herder organizations, followed by a chapter on drought management. Drought is a central concern in all arid zone pastoral systems, and supporting or amplifying existing drought strategies must be the first consideration once an institutional framework is in place.

Technical inputs are considered in Chapter 10, with emphasis on the *phasing* of inputs rather than offering technical prescriptions. The sequencing of inputs, so that they do not conflict or overburden local management capability, is one the most critical aspects in the introduction of technology, and there is now considerable experience on which to draw. The final chapter in this section concerns the design and operation of the monitoring system which all process projects need in order to adjust development inputs according to progress that is being made.

Users of this section should appreciate the speculative nature of pastoral development. Although there is much experience and improved understanding on which to draw, the packages now presented as 'good practices' are still unproven because there have yet to be pastoral development projects based on all that we now know.

7
Herder Organizations

Herder organizations have a central role in pastoral development. Considerable experience has now accumulated in World Bank projects, especially in West and North Africa. The West African experience was reviewed in World Bank Discussion Paper No. 175 on resource management and pastoral institution building in the Sahel (Box 7-1). But neither that discussion paper nor earlier papers supporting the concept of community-based organizations (e.g., Sandford 1981, 1983) devote much attention to the manner in which organizations must adjust according to the tasks expected of them.

The main thrust of this chapter is in general accord with the recent contribution by Swift (1994, Figures 7-1 and 7-2), but gives more emphasis to adjusting organizational forms according to the category of pastoral system and the functions that organizations are expected to fulfill. The chapter deals first with functions, before reviewing options and focusing on the main categories of organization that are likely to feature in future development.

Functions of Organizations

Herder organizations can be called upon to exercise NRM and fulfill a variety of service, marketing, and advocacy functions. Depending on the resource being managed or the other functions envisioned, organizational requirements vary.

NRM involves principally (a) managing grazing and water, (b) defending territory, and (c) negotiating and resolving conflicts over access to resources. An additional function is to hold devolved statutory rights and responsibilities in these fields.

Grazing management, while often pursued within a framework controlled from a higher level, is organized principally at the level of family units and neighborhood groups. Groups are involved when decisions are taken on major movements and reserving areas for later use. Different functions are involved when protecting grazing for use later in the year and when accessing grazing for use ahead of other users and before it deteriorates. The planting of fodder shrubs and trees (*Atriplex*, *Acacia*) can also be delegated to community organizations, as is happening in the Matruh Natural Resource Management Project in Egypt (Box 6-2).

Water management also involves several organizational levels, but with an additional distinction that surface water and shallow wells can be managed without heavy inputs from outside sources, while boreholes require fuel and other external inputs and often serve users from several groups.

Securing territorial rights is a matter of concern to all, although responsibility for preemptive or defensive action rests usually at the higher levels of social organization. There are also additional organizational distinctions that:

- exclusive customary rights are maintained through the institutions of one pastoral society;
- shared access rights are maintained through melding the institutions of the sharing groups; and
- statutory rights can be held only by persons or organizations entitled by law to hold such rights.

Conflict resolution must be exercised at the level appropriate to the conflict. Territorial conflict comprises (a) local disputes that can be resolved within a neighborhood group; (b) disputes that societal leaders can resolve; and (c) major incursions that can only be resolved (at least in favor of the weaker party) by State intervention. Disputes over water follow a similar pattern, although with the distinction that even local disputes (for example, over a single well) often require referral to higher authority.

Box 7-1. Pastoral Institution Building in the West African Sahel

A 1991 review on the experience of establishing pastoral organizations (POs) as instruments of NRM in Senegal, Mauritania, Mali, and Niger, and subsequent experience, led to several conclusions:

Rationale. POs can become instruments of decentralization and local development. In particular:

- at the international and national levels there is a new consensus on the need for community-based NRM institutions; and
- at the local level pastoralists realize that they have to come to terms with the State for economic and political reasons and therefore feel a need for POs and literate leaders.

Policies. There is continuing dominance of top-down procedures in PO formation. Also, land and water rights remain unresolved to varying degrees in all countries, in part because pastoralism is not seen as an activity that uses land efficiently.

Project lessons. Projects have not given high priority to creating enabling environments for PO development, have underrated the probability of drought, have been staffed predominantly by veterinary personnel, have lacked participation of women and social scientists, and have not clearly defined who the project was supposed to benefit.

Issues at the community level. Pastoralists give high priority to food and resource security and services such as water, health, credit, and literacy. Water points provide an opportunity to organize cohesive groups. Absence or low levels of literacy, management skills, and revenue generating abilities are major obstacles to institution building and sustainability.

Noteworthy achievements. POs have contributed to the growth of environmental awareness. They have contributed variously to dune stabilization, flood plain pasture rehabilitation, conflict resolution, animal health services, functional literacy, and increased political awareness.

Future progress. A shift is needed from the present project approach to an institutional program approach, guided by a long-term perspective with commitment by governments to promote improved NRM and pastoral development at the local level. Requirements include:

- an enabling environment for POs that implies institutional reforms based on decentralized tenurial control and pastoral institution building, strengthening human resource capacities, encouraging popular participation in planning and implementation, and establishing national-level coordination units and pastoral development centers at subproject level;
- increased budget allocations for PO formation, both within projects and for national pastoral institution-building programs, with an active role of pastoralists in revenue-generation and cost-sharing;
- extended national literacy programs and primary education, with adapted training for development agents and pastoral organization leaders and recognition of the role of women;
- better public health facilities and privatized animal health services where practical; and
- improved system of PO formation and NRM.

Decentralization. This is a key issue because dominance by central government in the creation of POs can only lead to suppression of local institutions and excessive dependence of pastoral groups on the public sector. Furthermore, future programs must ensure forms of organization geared to the specific functions expected of them.

Source: Shanmugaratnam *et al*. 1992.

Service functions have different organizational implications depending on the nature of the input.

Veterinary and medical supplies are widely used and yet require specialized procurement and handling. This implies one of two forms of organization (or both forms operating in parallel). One is a specialized commercial enterprise, self-governing and entrepreneurial, that retails supplies at prices that ensure a profit, and the other is a community-based entity that operates by subscription across a wide membership and employs a trained dispenser. If the organization is to procure in bulk, then it is likely to need letters of credit and so requires a form of organization that is recognized as a legal entity for that purpose.

Borehole operation calls for a trained operator and regular supplies of bulky fuel. Only where there is a cluster of boreholes is the supply of inputs likely to attract an entrepreneurial solution. More often a community-based approach would be required, perhaps based on a small core membership that retails water to other users.

Other services are often seasonal by nature. When services do not have to be organized on a regular day-by-day basis, responsibility is likely to devolve to an organization with another role that can be activated as required. This would apply, for example, to the organization of drought feed supplies.

Livestock marketing may involve all members of a community, but more usually will involve no more than half of all households and possibly about 20 percent. It makes sense for a marketing

Figure 7-1. Present institutional and organizational relationships in the administration of pastoral tenure

Geographic area	Types of management decisions	Present legal regime	Present key organizational actors: Customary	Present key organizational actors: Formal
Dispersion/ refuge territory	Negotiations on reciprocal access in bad years	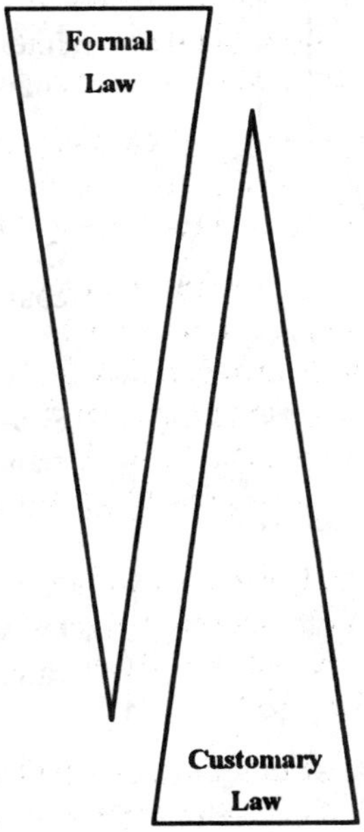	Inter-clan councils	Local government
Annual clan/lineage grazing territory	Access/ infrastructure development, major conflict resolution		Clan/lineage councils	Technical services
Dry season territorial base	Access, water infrastructure, conflict resolution		Sub-clan/lineage councils	Technical services
Key resources	Access, local investment/local conflict resolution		Groups of households/camps	Technical services
Individual/point resources	Access, investment, individual conflict resolution		Households	Technical services

Source: Swift 1994.

organization to draw its membership from those who are most likely to be selling livestock. It may also be appropriate to have separate organizations for different species if they are to be marketed to separate outlets at different times of year.

Advocacy and political representation require an organization that ensures input from those whose needs are being advocated, while leaving lobbying to those most likely to succeed (by power of persuasion and freedom of access to the right people at the right time). Such organizations are likely to be most effective when operating at the national or district level. National organizations now exist in Morocco, Mauritania, Niger, and the Central African Republic, and are starting to become important instruments in herder empowerment.

Multi-purpose organizations without a clearly defined role often run into difficulties, especially when governments dictate the form that local organizations should take. For example, those now operating in the Sahel of West Africa are all either *groupements d'intérêt économique,* or are classified by orientation as being *associative, mutualiste,* or *cooperative*, rather than in terms linked to customary concepts of scale and purpose. They operate — even those that have been in existence for twenty years — more as dependencies of government than as expressions of community purpose.

A further feature of the Sahelian organizations is that they are village-based, with no provision to ensure the rights of pastoralists engaged in transhumance. Indeed *groupements* are sometimes seen by village-based communities

Figure 7-2. Potential institutional and organizational relationships in the administration of pastoral tenure

Legal regime

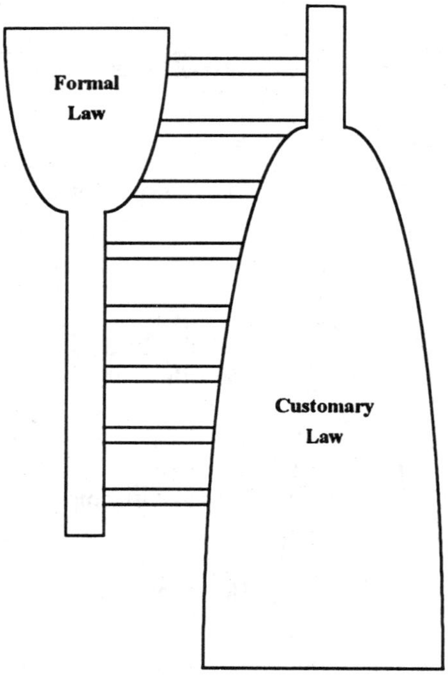

Geographic area	*Future key organizational actors*: *Customary*	*Mixed*	*Formal*
Dispersion/ refuge territory	Inter-clan councils	Local councils	Local government technical service
Annual clan/lineage grazing territory	Clan/lineage councils	Federations of pastoral associations	Enabling framework only
Dry season territorial base	Sub-clan/lineage councils	Pastoral associations/ committees	Enabling framework only
Key resources	Groups of households/ camps	Pastoral associations	Enabling framework only
Individual/point resources	Households		

Source: Swift 1994.

as a means of gaining precedence over more mobile pastoralists. Those so motivated then seem more inclined to follow official advice on NRM, while others focus more narrowly on infrastructure and services. In the case of service cooperatives in Ethiopia, interest focuses on shops and services rather than on NRM.

Establishing Herder Organizations

Ensuring appropriate herder organizations requires a three-step approach — first understanding existing social-territorial organization, then formulating development objectives and priorities, and finally identifying and assisting forms of association that build appropriately on customary institutions.

Social-territorial organization raises two points important in the present context.

Decisionmaking needs to be thoroughly understood. There is a tendency to identify the principal organizational unit and then to load that one structure with all responsibilities envisioned under the project. Instead, it is necessary to establish how individual functions in NRM are exercised and then to proportion responsibilities and strengthen local management capability accordingly.

Resource tenure warrants particular attention. If land is being lost from the pastoral system, upgrading NRM capability is not going to be of much help. If local institutions are proposed for legal recognition in order to counter incursion, then it is necessary to ensure a form of organization that the law can recognize as competent to exercise authority over land.

Development objectives and priorities should be set out in the form of a development path. The immediate concern is to decide what goes into the first phase of operations, and the process approach will then take care of what follows. Three points are important.

Institutional capacity should be allowed to guide project content. Inputs that are thought to be desirable but do not have an existing mechanism for their management or control probably do not belong in the first phase of the project at all.

Organizational support needs to be provided at two or three levels, one or two relating to specific functions such as animal health or water management, and another supporting the leadership structure, not necessarily in terms of any physical input, but to ensure official recognition so that lower-level organizations can receive customary support in conflict resolution.

Subsidiary users often need to be accommodated to ensure that seasonal rights of access are honored by primary project beneficiaries or are met in other ways.

Evaluating organizational options relies on being able to differentiate organizations in terms of their principal attributes. Table 7-1 is a check-list of the six most basic attributes that must be considered in order to assess the competence of an organization in any particular context.

Apparently innocuous features can be highly influential. An inappropriate constitution, for example, can cripple an organization that holds far-reaching responsibilities. More information on constitutions and links between organizations and security of land tenure is included in the next section.

Categories of Organization

Although the variables listed in Table 7-1 potentially lead to a great variety of organizational types, the forms of organization likely to be used can fit within a framework of five categories. Each has its own character that fits for some tasks and not for others. Some can also fit within a hierarchy of organizations — an arrangement that often needs to be adopted in order to ensure coordination among several activities.

The five categories are described under titles that provide a best fit in plain English, while avoiding terms with a specific meaning in law such as 'group' in Kenya. The term group or grouping is used here, as elsewhere in this paper, to describe sets of families who are linked socially and geographically but in no other specified way.

Committees are groups of people self-selected or appointed by decision of a larger body for purposes of reviewing or overseeing a matter of common interest. They may be action-oriented decisionmaking bodies and able to commit resources, but they have no status in law and so cannot own property or be held legally responsible for their actions. Membership seldom exceeds twenty, and they work to a simple constitution or terms of reference.

Committees can be differentiated on the grounds of being permanent ('standing') or *ad hoc*; high-level ('supreme') or local; or traditional or 'modern' (usually government-inspired). Obviously they can also be differentiated according to their function. A standing local committee (traditional or modern) might exercise management responsibility for a village well, while an *ad hoc* supreme traditional committee might be convened for inter-tribal discussion on a new aspect of government policy. Permanent authoritative committees, however, are classified below.

Councils are permanent and authoritative decisionmaking bodies that derive their authority either from a statutory source or from societal

Table 7-1. Checklist of Key Attributes to Differentiate Organizations

Attribute	*Significance*	*Principal categories to differentiate*
Affiliation	Differentiates organizations as being:	• prescribed by (a) central or (b) local government; or • an indigenous structure, either (a) traditional or (b) "modern" (expressing popular grass-roots aspirations).
Representation	Characterizes the membership in terms of:	• ethnicity, whether (a) unified by language and *mores* or (b) involving two or more disparate groups; and • qualification for entry, whether by (a) residence, (b) common interest, or (c) social or economic status (and specifying criteria and restrictions by age/gender).[a]
Purpose	Specifies whether the primary motivation is:	• social advancement or security, whether through (a) political advocacy, (b) territorial defense (by land title or force of arms), or (c) organizing social welfare measures; or • NRM or economic benefit, specifying (a) resources managed, (b) services delivered or (c) commodities involved (produced or marketed).
Size	Distinguishes whether the organization is:	• localized (e.g., up to district level), differentiating membership as (a) <40, (b) 40-200, or (c) >200; or • extensive (regional or national), differentiating membership as (a) <200, (b) 200-1,000, or (c) >1,000.
Governance	Differentiates organizations in terms of:	• officers (a) elected by democratic process or (b) appointed by 'higher authority'; and • the written constitution (a) formulated by majority decision or (b) furnished by the State.
Legal status	Differentiates organizations as being:	• a legal body corporate, or • without legal status, either (a) 'registered' (conferring quasi-legal status) or (b) 'informal' (with no State recognition).

a. In the case of established organizations, it is also useful to quantify membership as a proportion of those eligible since an organization that has attracted the majority of its potential membership is clearly in a more effective position than one that represents a minority interest.

convention. They usually oversee a range of activities and are often valuable in project coordination. Statutory and traditional councils, with their accompanying executive structures, can be distinguished as follows:

Statutory councils are either government structures with a mandate for sectoral or area development, or they are the instrument of local government. They may have pastoralists as members, but seldom practicing herders, and so are not herder organizations.

Traditional councils and their leadership structures maintain societal values and are the custodians of customary law. One feature of leadership in pastoral societies is that it usually embodies both vertical and lateral extensions, reaching down to the neighborhood level and sideways to recognize the authority of specific individuals in designated areas of religion and custom. Although governments are often antipathetic to traditional leadership, these structures cannot be ignored in project formulation and management. They are often the best rallying point for NRM at the neighborhood level. At higher levels it may be sufficient to ensure their role in conflict resolution and perhaps to codify unwritten rules of conduct. Leadership structures can possibly be used in a project context to help ensure liaison within a hierarchy of organizations, but it is more likely that the project will need to set up a committee of its own for this purpose, including government officials.

Associations are common interest groupings, either traditional or modern, that provide a framework for concerted action by a substantial body of people. Associations hold no statutory responsibility for land or property, although if registered with a government department or local authority, they may enjoy quasi-legal status sufficient to allow them access to credit (and to rights in property other than land if registered as a society). In such cases they would have a written constitution provided or approved by the overseeing authority. Traditional associations seldom have (or start with) a written constitution, although their functions and procedures are usually prescribed by customary law.

Associations should be differentiated as registered or informal and according to their role in management or advocacy. Management responsibilities could be specific to grazing resources, water supply (such as a borehole), other services, or marketing. Although some of these functions could be combined as the responsibility of one association, focus and effectiveness may be lost. When introducing new associations to build management responsibility and capability in unaccustomed fields, it is best to focus initially on the inputs and activities in greatest demand (for example, new water supplies or animal health services). Associations should have a relatively homogeneous membership. A typical association might comprise 20 to 200 families.

Corporations are legally constituted 'bodies corporate', with right to own land and other property and to sue and be sued. The normal type of registered society does not qualify on these grounds, but cooperative societies and companies do, as do groups in Kenya with registered land rights. Invariably these organizations are known by their corporate name (as companies or cooperatives, etc.), so the corporation title is seldom used.

There are legal requirements covering registration, adoption of a constitution, election of officeholders, keeping accounts, and procedures for dissolution. These requirements make corporations unwieldy vehicles for pastoral development. Some form of legal corporate status is necessary, however, where secure tenure calls for land title or where the community concerned wishes to engage in business (such as hiring contractors or taking a substantial loan for water development). If pastoralists and planners are adamant that corporate status is needed and yet the forms prescribed by law are all too complicated, then it may be appropriate to legislate for a new form.

Corporations can be differentiated as (a) freehold or leasehold (or untitled if they hold no land), (b) commercial or subsistence, or (c) production, marketing, or service oriented. There is usually latitude within the prescribed legal framework for corporations to adopt their own constitution and procedures in matters relating to specific corporate objectives. Much hinges on using this provision wisely, and corporations should be encouraged to stick as close as possible to customary provisions with which they are familiar. If the legal framework is too onerous, a project may be able to use its influence to amend the law. In any event, officeholders need counseling.

Federations are alliances of associations or corporations formed for overall grazing management or for sharing resources or joint ventures such as water development. This category will certainly find a place in future pastoral development. The assumption is that participating bodies would retain their existing status and that federations would opt for whichever status best suits their individual functions. The constitution that they adopt — particularly for governance and provisions for conflict resolution — would be critical to their effectiveness.

Legal Implications

If customary institutions or organizations are to be given legal recognition, then it is necessary to review which institutions are to be 'upgraded' and what government institutions are needed to oversee and assist in the process.

Customary institutions need careful review because they do not now always work for the benefit of their constituency at large. If land title is to be granted through a process of statutory adjudication, then the outcome will be determined by how the law is written and how pastoral groups present their customary rights for adjudication. Otherwise recognition of land rights is open to negotiation, and is likely to involve one or two levels:

Societal recognition would arise if the State were content to recognize the territorial rights of complete pastoral societies at clan or tribal level. This could be done without necessarily confirming or denying rights existing below that level. Societal recognition is the most direct and simple means of safeguarding pastoral land rights, provided that the customary tenure system is competent to manage affairs internally. It also has the advantage of formally recognizing customary leadership structures that can be critical to effective pastoral development. A variant would be to recognize rights only in homelands or dry-season grazing areas, while leaving wet-season areas unallocated and open for negotiated access.

Local group rights can be confirmed within or instead of societal recognition. This approach applies only where there is a relatively tight form of social-territorial organization within which the rights of individual clans or other groups can be recognized. The smaller the units, the more necessary it is to attach provisions for reciprocal access. The option of recognizing local group rights while withholding societal recognition could be used as a ruse to exclude the leadership structure, or it could be the means to bypass a customary tenure system that has become badly corrupted.

Government institutions need to provide intermediaries to begin confirmation of rights and to provide longer term inputs to registration of rights, monitoring, and counseling.

Intermediaries are needed to introduce pastoralists to the procedures available for improving security of tenure. This could be the role of the district administration or of a pastoral development unit, depending on the structure of government. If allocation of freehold title is an option within the law, then intermediaries have the extra task of ensuring that pastoralists are aware of the need to present themselves for adjudication in groupings that will pass adjudication as being 'customary', while also combining social cohesion and ecological integrity.[1]

Registration of land rights is necessary to counter disputes. Registration is usually best arranged at district level because centralized registration can become remote and cumbersome. If there is a legal requirement for a central

1. *Freehold title* has disadvantages (see under group ranching in Annex B), but if it is an entitlement that is open to pastoralists, the choice should be theirs and should not be made paternalistically for them by officials or development advisers.

registry, it may still be possible to designate an existing cadre of officials at the district level to exercise the registrar's functions. Bear in mind also that recording reciprocal rights can be just as useful as registering principal rights.

Counseling will be discussed in the next chapter as a function that local government should extend to newly-formed herder organizations. The job is similar to that described for intermediaries, but calls for more intimate knowledge of the organizations concerned.

8

Support Systems

The surest route to pastoral development is through strengthening pastoral institutions and minimizing dependence on external inputs. Nonetheless, external support is needed. The traditional approach has been to strengthen the capacity of government ministries to deliver such support, but it is now evident that public budgets cannot cover all requirements. The present trend is toward cost recovery and devolving service delivery to the private sector, with central government focusing on policy and control and providing essential infrastructure.

This chapter examines the contributions that should come from central government, local government, and the private sector. Seven areas of support for development are considered (Table 8-1). As might be expected, there are several areas of responsibility that cannot be devolved and so remain with central government.

Central Government

Policy frameworks establish the level and nature of development support that is provided, including the role of contributors other than central government. The policies cannot just be restatements of national goals and aspirations — for policies to have meaning in pastoral areas, they need to be grounded in concepts of sustainable resource use. Resource use strategies, in turn, must reflect the realities of population pressure, capacities for cost-sharing, and the values attributed to rangeland resources, including regard for the ecology, water resources, and pastoral systems of the area. These strategies establish the basic forms of resource use and social-territorial organization that projects need to preserve. The need is most pronounced where the choice is between preserving seasonal movement or presiding over a descent into sedentary squalor. The final step is to ensure enabling legislation as the bottom line in creating a policy environment supportive of pastoral development.

Projects should contribute where needed to all of the foregoing, including any field research needed to define appropriate resource use strategies. This is not a call for open-ended research, but for specific information on key aspects that should determine the direction for project intervention:

Population pressure and poverty levels need to be interpreted in terms of whether 'development' means optimization based on present populations, or seeking redeployment into other areas or occupations.

Capacity for cost sharing needs to be established for the pastoral economy as a whole and for each social stratum represented. Most projects proceed on the basis of a rather hazy notion of the ability of households to contribute, although this information is basic to knowing if subsidies are a policy issue and if the private sector has a role.

Rangeland values need to be established in terms of both present-day pastoralism and wider- and longer-term perspectives (Chapter 3). Sometimes future values cannot be predicted (as when there is an undiscovered mineral deposit), but wildlife potential can usually be evaluated to decide if there is a case for moving closer to the natural state.

Basic infrastructure usually needs to be improved to enable support services to function efficiently. Three points deserve attention during project preparation:

Physical infrastructure, like policy, should support what is strategically important. It is more cost-effective and less disruptive environmentally to plan roads and other facilities to serve a resource use strategy, rather than to let other criteria determine where facilities are sited.

Service centers are needed to bring technical and social services closer to people. Their siting should consider who is to run them. Local government or herder organizations may be better able to keep remote centers staffed and provisioned than central government, which often has difficulty maintaining centers far removed from district headquarters.

The institutional framework for development support derives from the policy

Table 8-1. Development Support for Natural Resource Management

Support area	*Activity*	*Primary responsibility*
Policy framework	Research to assess: • population pressure, • capacity of pastoralists for cost-sharing, • value of range resources Leading to: • strategy for resource use, • policy statements, • enabling legislation	Central government (with the possibility of contracting out studies)
Environment protection	Inputs with environmental objectives, e.g., • tree planting, • alternative fuels/stoves, • fire-control, • wildlife management	Local government, implementing central policy and encouraging private sector input where feasible
Drought management	Inputs to counter low and erratic rainfall: • early warning systems, • contingency measures, ◆ strategic livestock offtake (and banking) ◆ food-for-work ◆ grain reserves • recovery measures	Central/local government with the possibility of inputs by traders and banks
Basic infrastructure	Physical infrastructure • road network, • other basic facilities Institutional framework	Central/local government
Social services	Health services Primary education Adult literacy	Central/local government, with inputs by NGOs
Technical services	Extension Credit Animal health services Water supply systems Marketing support Agronomic inputs (especially in agropastoral systems)	Private sector (through subcontracting and subsidies from central government), with local government becoming increasingly involved in backstopping
Institution building	Counseling/training (aimed at herder organizations and helping pastoralists plan, operate, and monitor development activities)	Local government, with increasing contributions from national and regional herder federations

framework already discussed. Unless central government delegates a role for local government and adopts fiscal policies conducive to private sector involvement, neither will be able to participate in the manner suggested later.

Other inputs expected of central government depend on the extent of devolution just noted. Some specific requirements are noted as they arise in the sections following and in the chapter on drought management (Chapter 9). Basically, the role of central government is to encourage local initiative, ensure necessary infrastructure, and mobilize adequate inputs in times of need, including, in the case of drought management, famine relief goods from outside the country. Central government also retains responsibility for ensuring quality control in all areas delegated to others and in the control of insurgency. Although traditional leadership structures could play a more active role in conflict resolution (Chapter 7), situations arise periodically where conflict is beyond societal control, particularly where incursion comes from outside the country. Often governments are reluctant to admit incursion and insurgency as a problem, but such situations need to be identified since conditions of insecurity can nullify all development efforts.

Local Government

Shifting responsibilities from central to local government fits well with current development concepts of decentralization, subsidiarity, and democratization. Although national governments are becoming more democratic, pastoralists often remain far removed — geographically, linguistically, culturally, academically, and economically — from those who run the country. Pastoralists can usually identify better with their locally elected representatives. On the other hand, local governments usually have limited resources and are less efficient than central government, so improving their competence implies seconding staff from central ministries, where most trained personnel are presently concentrated. This would improve accountability — by bringing ministerial staff under the day-to-day control of elected representatives of the people they are serving — but the whole idea is usually unpopular with both central authorities and the technical staff.

As a compromise, project planners turn to NGOs to deliver services to pastoralists, but this does little to achieve longer-term sustainability. The preferable approach may therefore be to involve local government in selected support activities to which they are particularly suited, such as:

- institution building in herder organizations,
- contributing to drought management, and
- inputs to environmental protection and social services.

Institution building has several dimensions. The policy framework established by central government will determine how much recognition is extended to customary institutions and organizations, but local government has a role in promoting the decisionmaking process. Thereafter, local government should be involved, either directly or by guiding the inputs of others, in building the capabilities of herder organizations by:

- counseling on organizational mandates and constitutions;
- advising on how to adapt customary procedures to new tasks;
- training in tasks relevant to each organization (covering options open to the organization, recordkeeping, contract services, operating health services, boreholes, etc.);
- brokering necessary social and technical services; and
- ensuring inputs to drought management and monitoring.

Drought management offers a pivotal role for local authorities, by providing a coordinating office set between the pastoralists who are operating their front line strategies and the central government that is fulfilling the role already described. It is at this middle level that grain stocks and other drought emergency reserves can most conveniently be held.

Environmental protection is another area in which local government can help bridge national and local perspectives. Tree nurseries and tree planting are best organized at district level. Control of overcutting through tree planting or provision of new energy systems can readily serve both interests and warrant a place in middle-level support systems. Not too much should be expected of alternative fuels and new devices in pastoral systems where wood fires have multiple uses as a social focal point and deterrent to predators, and for roasting whole carcasses.

Wildlife management schemes are also sometimes of mutual interest. Such schemes are commonly conceived centrally but then operated through local authorities. This is how the Communal Areas Management Programme for Indigenous Resources (CAMPFIRE) started in Zimbabwe. Although a dispute has arisen over whether revenue from safari and trophy fees should go to Rural District Councils or directly to the households concerned, CAMPFIRE is still a model on which other schemes can build (World Bank 1996).

Social services also benefit from strong local involvement even though the overall direction is centralized. One advantage of local government structures is that they are not forced to think sectorally. For example, where ministries have difficulty developing unified delivery systems, a local council can readily train and equip auxiliaries to deliver primary health care for both people and livestock.

Private Sector

Many factors conspire against private sector involvement in pastoral development. Distances are great, people and markets are scattered, and there is usually more money to be made by trading elsewhere. Nonetheless, there is potential in some lines of business, especially if herder organizations include service cooperatives that can act as intermediary between pastoralists and suppliers. Technical services, although now mainly the preserve of government departments, are worth examining for potential private sector involvement. The best prospects lie in specific inputs that combine low volume with regular demand.

Animal health services are the prime example. Government still must ensure an enabling environment — extending in this case to permitting importation and trade in veterinary products — but then herder associations or corporations can provide the economies of scale needed for primary suppliers to include pastoral areas among their outlets. The low productivity and density of the livestock population usually precludes private professional practices, but community health assistants (paravets) can be trained and backstopped by a few professional veterinarians. Sub-contracting of 'public good' tasks such as compulsory vaccinations to private veterinarians, as in Morocco (Box 8-1), is a good way to encourage private veterinary practice in low-potential environments.

Other services need to be assessed individually within each project for potential delivery by the private sector. Individual households in pastoral systems vary greatly in their capacity to pay a commercial price for services that they want.

Water supply systems offer varying prospects for private sector involvement. Drilling for water and excavating ponds attracts contractors, but maintenance is more difficult to assure. Federations would find it easier than smaller organizations to negotiate competitively-priced maintenance contracts.

Agronomic inputs for agropastoral areas could in many instances be delivered by the private sector. Accelerating a switch to private sector delivery may not make much sense, however, if technical assistance beyond private sector capability is needed to ensure effective use of the technologies being introduced.

Credit and marketing support are additional areas for review. Private sector involvement through commercial banks and traders often represents a better option than direct government involvement. Because credit and marketing are particularly important in the context of marketing for drought management, they are considered further in the next chapter.

Box 8-1. Sub-Contracting Compulsory Vaccinations to Private Veterinarians in Morocco

The Moroccan Livestock Service began to privatize veterinary services in 1983. Since then the number of private veterinarians has increased from 2 to more than 200, about half of the country total. Today private veterinarians care for 50 percent of the cattle and 40 percent of the sheep in the country. The continuing factors for this success are:

- the political will of the Livestock Service to make privatization a success;
- the immediate suspension of curative services and non-compulsory vaccinations by the Livestock Service once a private veterinarian has established a practice in an area;
- a well-functioning national association of veterinarians; and
- a clear subcontracting policy for compulsory vaccinations, leading to a higher proportion of livestock being covered at 50 percent lower cost (Umali, Feder, and de Haan 1992).

Source: World Bank 1994.

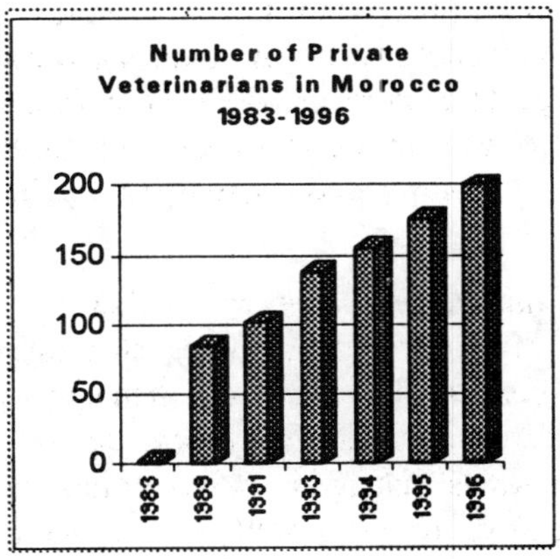

Source: MAMVA (Ministère de l'agriculture et de la mise en valeur agricole, Direction de l'élevage). Personal communication.

9

Drought Management

Drought is a natural feature of arid areas, yet projects often treat it as external to pastoralism, covered by sensitivity analysis as though drought were a freak of nature rather than the surest feature of the project environment. Projects need to be designed with drought management permeating all components, or with a separately identified drought management component. Such a component should support customary drought strategies, improve drought management capabilities, and provide drought recovery assistance. The starting point is to agree on a definition of drought.

Definition

Drought is a condition of unexpectedly low rainfall — an important criterion because there is a tendency to invoke drought whenever grass or crops fail and people suffer, even when rainfall is above average. The suffering is real, but its treatment will not be helped by misdiagnosis of the cause.

What constitutes drought, in terms of rainfall, can be established only by analysis of long-term rainfall data to appreciate what is 'normal' for an area. Although past performance provides no guarantee of the future, it is irresponsible to ignore past patterns. The essence of drought management is to ensure a capability to respond to drought episodes of different intensity. There is no excuse for being surprised by the only sure element of arid zone pastoralism.

Drought intensity must be interpreted relative to the status of a pastoral system. A healthy pastoral system should be able to treat as 'normal' years in which (barring very freakish distribution) rainfall is 75 percent of the long-term average. Drought should only need to enter the vocabulary of pastoralism when seasonal rainfall lapses to about half of the long-term average, or where two or more years pass without exceeding 75 percent of the average. To be realistic, however, with increasing population pressure, it is now quite usual to invoke drought when rainfall is 150 percent of average.

Table 9-1 illustrates this relationship. The table should not be taken too literally, but it does help to identify three intensities of drought against which to consider how strategies might be brought into play.

Table 9-1. The Relation Between Rainfall (average = 100) and the Duration and Severity of Drought as Affected by the Status of the Pastoral System

	'Healthy' system		*Depleted system*	
Severity of drought	*1-2 consecutive poor seasons*	*2-4 consecutive poor seasons*	*1-2 consecutive poor seasons*	*2-4 consecutive poor seasons*
Mild	<75[a]	<100	<125	<150
Average	<50	<75	<100	<125
Acute	<25	<50	<75	<100

a. For example, rainfall is less than 75% of average, but greater than 50% of the average.

Mild drought can usually be borne by using up grazing reserves and perhaps selling more livestock than would normally be the case. As drought lengthens, additional livestock would be sold and herders would dig deeper into available reserves, especially browse resources. Movement patterns would not be much affected, nor should famine relief be needed.

Average drought begins to affect customary movement by changing transhumance schedules or detaining animals in their dry-season grazing area. The last of any preemptive livestock sales would take place. Remaining reserves of grass and browse would be used up, perhaps aided by deepening wells or otherwise providing water in outlying areas. Famine relief would be needed, first by poor semi-sedentary families and then more widely as drought lengthens.

Acute drought would first be accommodated as indicated above, and by migration into drought retreats if these are available. If migration were not feasible, there would be increasing concentration around reliable water points (often in the vicinity of district headquarters) and increasing dependence on famine relief.

Although 'depleted' systems can be treated in the manner indicated, they really need more fundamental help since they fall into the first two situations described in the section on environmental sustainability in Chapter 6. Nor does Table 9-1 distinguish between the circumstances of the drier and wetter cycles mentioned in Chapter 1. If there is reason to believe that a dry cycle is starting, a more considered response than the 'one step at a time' scenario implied above is necessary. For example, provision for the 'average' responses might be put in place permanently.

Analysis of Drought Strategies

The basic strategies used by pastoralists to counter drought have been indicated for three intensities of drought. Drought is either absorbed, avoided by moving elsewhere, or under certain circumstances, dealt with by disposing of livestock and turning temporarily to other sources of subsistence. There are many different ways to absorb drought. It helps to start with hardy, drought-tolerant livestock and detailed knowledge of the vegetation so that the best use can be made of available resources.

In principle, drought resilience could be gained by switching to more drought-tolerant breeds as the risk or intensity of drought increases, but this degree of flexibility is seldom practical. A pastoral society that is cattle-based tends to favor cattle even when living in an environment that is better suited to camels and goats. However, pastoralists commonly change the relative proportions of the species in their holdings, and may turn to 'new' species as an adaptation to radical change in the accustomed environment.

The effectiveness of strategies depends in part on their timing, whether applied as a preemptive or reactive response to drought. Reactive responses are more usual in pastoral situations, except when movement has become so institutionalized that people move automatically at a time when there is not a current threat to life or livelihood. Otherwise, the response tends to be left until the last minute, when possibly fewer options are open than would have been the case had action been taken sooner. It is often a project objective to replace some of these last minute decisions by preemptive strategies.

Migrating to avoid drought is a characteristic pastoral strategy. Most forms of transhumance incorporate this principle and can achieve added drought security simply by adjusting the timing of movements. Less mobile systems would also have used migration to avoid drought at times of real need, although these days there are fewer places where pastoralists can go to escape drought. Only pastoralists whose acknowledged territory extends across a wide ecological spectrum can now use mobility as their principal drought strategy.

Relying on local grazing reserves without long-range movement serves only to cope with mild drought. The practicality of this reliance depends on area, ecology, and population pressure, and works best if applied preemptively by setting aside areas for later use and using the remaining areas more intensively or efficiently. Part of the 'reserve' can be coarse grass or relatively unpalatable browse that is used when there is nothing else available. The technology of adding polyethylene glycol ('browse plus') to stock water to increase the intake and utilization of browse is still at the testing stage under ranching conditions and is not yet available for pastoral areas.

Preemptive sale of livestock in advance of drought — before their condition worsens and value decreases — makes sense economically, and also makes it easier to nurture the livestock that remain. This sale presupposes, however, that an effective banking or credit system is available to facilitate restocking when conditions improve. It also requires that grain foods be available at a fair price to fill the dietary gap left when livestock are sold. Where these conditions are met, the strategy is worth promoting, at least in pastoral systems already trade-dependent or semi-sedentary. Its introduction elsewhere would encourage sedentarization unless monetarization of the economy leads instead to motorized herding.

Buying feed during a drought another way to mitigate the situation. It is not unusual for ranchers or settled pastoralists to buy hay or concentrate feed and whatever is needed for family subsistence. Governments also often take the same approach when helping their pastoral constituency through free famine relief or the supply of subsidized feed. The supply of subsidized feed seldom has the desired effect:

Equity imbalances arise because usually it is the most powerful (and therefore wealthier) herders who acquire the largest share of subsidized feed. Little quantitative information is available on the equity aspects of subsidized feed distribution systems because the topic is politically sensitive and there is not much experience on which to base the targeting of inputs to less well-off herders. Possible strategies to be tested include providing feed supplies only to smallholder feedlots, packaging and transporting feed in small quantities, ensuring strong community involvement in the organization of the distribution, voucher systems, etc.

Impacts on rangeland arise when imported feed leads to more animals being maintained on the range, not only during drought, but also after the first rains arrive. At that critical stage of recuperation, range vegetation is subjected to a stocking rate far higher than would occur if the drought had led (whether by death or sale) to fewer animals. This extra grazing pressure jeopardizes the survival of perennial plants and the capacity of annuals to flower and replenish seed stocks. The art of managing 'non-equilibrium' situations is rapid destocking in times of drought, and restocking once the rains return.

Long-term effects on pastoral economies occur when, as in many North African and Middle Eastern countries, the provision of subsidized feed becomes structural, as noted in Chapter 4. This use of feed increases or maintains stock numbers at consistently high levels. Figure 9-1 shows this effect on small ruminant numbers during a twelve-year period in Morocco (without any consistent change in range management). Where subsidized feed is less readily available, such as in sub-Saharan Africa, livestock must be sold to obtain cash for animal feed. The key issue is timing — as a drought progresses, livestock sales are hindered by low demand and depressed livestock prices. Figure 9-2 shows the effect of drought on cattle and grain prices in West Africa (Blackburn *et al.* 1993).

Packaging Project Assistance

The assistance that a project gives to individual strategies will vary greatly among pastoral systems, but the framework needed to provide overall support for drought management remains fairly constant. There are three basic requirements: (a) early warning to inform everyone of impending drought, (b) drought response mechanisms to ensure timely government responses, and (c) drought recovery measures. Box 9-1 describes the Emergency Drought Recovery and Arid Lands Projects in Kenya.

Early warning systems follow a standard form, whether monitoring a natural system or a production line — analyzing a flow of relevant data for indications of a critical state. Several countries already operate a drought warning system, but with a tendency to rely on remote sensing more than on data collected in the field.

Remote sensing implies using satellite imagery and the digital data from which the imagery derives. Whereas air photography and aerial survey are too expensive to commission at monthly or fortnightly intervals (the frequency required), imagery is readily available and even real-time weather data can now be received directly on-site for the cost of a satellite dish and a terminal. The interpretation of satellite data is still a specialized field, however, and is best done nationally or regionally and then disseminated as required. Because imagery is not particularly informative about ground conditions when vegetation is sparse and desiccated, its main use is to monitor the lapse of vegetation at the onset of drought and to track weather systems both then and throughout the dry season. Imagery is best used as a supplement to field monitoring.

Field monitoring is essential in order to maintain a monthly flow of information about the availability of grazing and water and the general state of crop and livestock production. Useful production parameters include trends in marketing (particularly the balance of trade between livestock and grain foods) and household parameters relating principally to diet, activities, and dependency across a social cross-section, including poor families.

Figure 9-1. Rainfall and Distribution of Feed and Small Ruminants in Morocco, 1980-1991

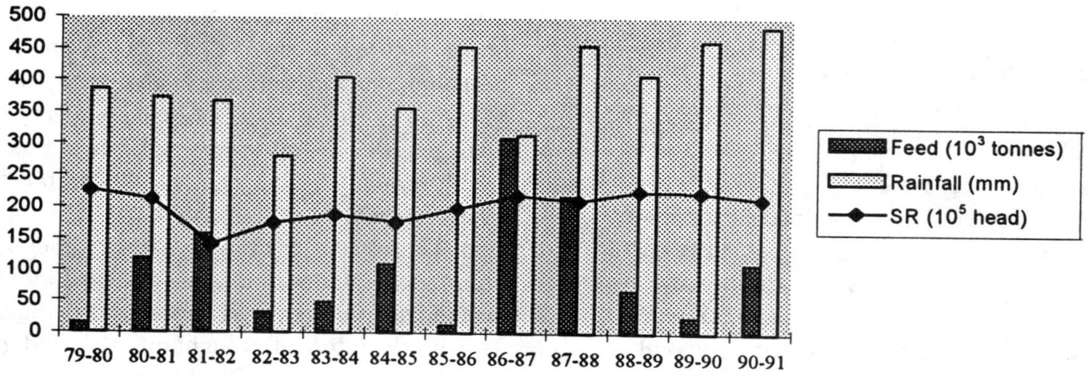

Source: Internal World Bank documents.

Figure 9-2. Effect of Drought on Cattle and Grain Prices in West Africa

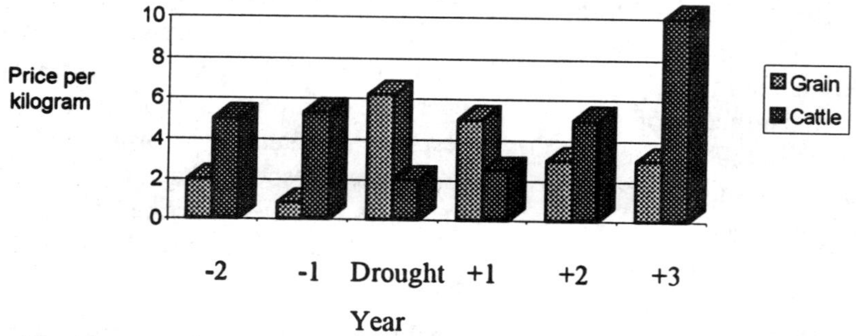

Source: Blackburn *et al.* 1993.

Maintaining awareness of these points need not be onerous, and much of the information can be furnished by herder organizations or village groups. Indeed, it is better to delegate the responsibility than to set up an independent monitoring team since it then gives the community a genuine participatory role.

Coordination of these activities involves three levels:

- District coordination is pivotal because it is at this level that key parameters and participants can best be identified. Also at this level decisions must be made about how to use the information most beneficially. Lead responsibility might best go to local government, with others asked to join a coordinating committee, and with technical assistance to get the process started.
- Local coordination is needed to collect and consolidate information at the subdistrict level. This reduces travel and is particularly important when a drought is in progress so that a prioritized picture of needs is forwarded to higher authority. If responsibility at this level does not rest with local government, it might usefully be held by the traditional leadership.
- Central coordination is needed to disseminate remote sensing data and ensure a degree of standardization in field data collection. Standardization helps central government to prioritize drought relief when required.

More important than data collection is the use made of the data. Because the response mechanisms described next are also multilayered, it helps to involve the same parties in both functions. Interest and commitment are heightened if those who collect the data are also involved in the use of those data.

Drought response mechanisms are built upon three components. Each must be assured individually and then melded into an operational whole. The aim should be to arrive eventually at a mechanism that is activated as herder

Box 9-1. Drought Recovery and Preparedness in Kenya — the Emergency Drought Recovery and Arid Lands Projects

The Kenya Arid Lands Project (ALP) is intended to institutionalize at the national and district levels a structure to effectively manage all phases of drought (preparedness, mitigation, and recovery). The project would implement a Drought Monitoring System (as tested under the previous Emergency Drought Recovery Project), establish a Drought Contingency Fund for immediate drought interventions, and an Emergency Cereal Reserve to maintain grain stocks at the district level. The project would also continue drought recovery activities for people made destitute by the 1993 drought, and finance other drought preparedness and mitigation interventions such as establishing dry-season and drought-grazing reserves and strategic water supplies.

The ALP could be a model for other arid land development efforts. The project follows a strongly decentralized approach, with responsibility for the planning and implementation of most activities transferred to district committees of local authorities and beneficiaries, and the operation of key pastoral input services, such as management of water points and distribution of veterinary pharmaceuticals transferred to pastoral associations.

A major challenge for the future of ALP will be to develop new drought preparedness approaches and strategies — including the identification of appropriate baseline information and monitoring indicators, promotion of regional pastoral associations, and further analytical work on livestock marketing as a key factor in drought management. More attention will also focus on exchange of information between West and East African efforts in the area of arid lands development.

Source: World Bank internal documents

organizations — with growing competence and confidence — decide what is necessary, with a response office at the district level and established procedures for mobilizing support from other sources. However, the components must first be established:

Scales of 'drought alert' need to be established, ranging from a baseline (situation normal) to increasing levels of threat and hardship. In effect, one preliminary stage of alert is needed, followed by others that match or slightly anticipate the three drought situations noted in the previous section. Emphasis should be on defining the early stage of alert that triggers the convening of a district drought committee. Thereafter, that committee can decide for itself when further significant deterioration has taken place. Nonetheless, the indicators of worsening drought need to be agreed.

Response schedules should be decided for each stage of alert. For the early stages, specifications should include: (a) the stage at which the district drought committee is to be convened; (b) the specific measures of preparedness that need to be reviewed (such as grain stocks, the state of reserve grazing, emergency watering points, and livestock marketing facilities, etc.); and (c) the stage at which central government and other interested parties are informed so that they have time to prepare their possible response.

Obviously the membership of the district drought committee must also be decided, along with matters relating to the logistics of its operation. Indicative bills of quantity need to be prepared for the action attached to each stage of drought intensity.

Means must be ensured to allow each party to make the response expected of it at each level of alert. This probably implies establishing a contingency fund of a size that provides, for example, 50 percent of the maximum likely expenditure. It is clearly impracticable to tie up a very large sum of money, and yet there must be sufficient to make a substantial emergency response. Indicative budget categories are:

- strategic reserves of famine relief;
- the medical and veterinary supplies needed to help people and livestock overcome the debilitating effects of drought;
- cash or credit to encourage livestock owners to offer animals for preemptive sale early in the drought;
- mobile abattoirs to allow field processing of remaining drought-stricken livestock;
- portable generators and tankers for lifting or moving water where concentrations of people make unaccustomed demands on water supplies;
- food or cash for work; and
- necessary transport and fuel.

Apart from money, strategic supplies of essential items need to be held in stock. There is no point in having a drought response capacity that takes several weeks to mobilize while essential items are imported. The whole point of drought preparedness is that it reduces the waiting (or dying) time.

Tactics to ensure timely response might include (a) building up reserve food stocks from seasons of plenty; (b) holding back part of the total development budget for release only in the last quarter of each financial year (after it is seen that no emergency situation is intervening); (c) establishing a mechanism for public subscription that can be activated at short notice; and/or (d) designating, equipping, and training selected military units for logistical support during drought emergencies.

Drought recovery is as important as drought mitigation. Some degree of natural recovery can be expected because surviving livestock populations show high natural fertility in the first years after drought. There is a 'pent-up' fertility in these populations that expresses itself in high calving and lambing rates once rains return. However, project-led rehabilitation is still often necessary.

The main component is restocking to enable herders to reconstitute their herds, either using money or credit derived from preemptive sales to buy back livestock, or restocking destitute families in the manner often practiced by NGOs. There may also be need for rangeland rehabilitation, but it should not be assumed that reseeding of depleted rangeland is always needed (see next chapter).

Buy-back requires supplies of livestock for purchase (especially young breeding stock) and buyers with the means to purchase. The livestock can normally come through standard marketing channels, with no exceptional input beyond the extra effort needed to draw livestock into the marketing system if the drought has been widespread. Ensuring that prospective purchasers have the means to purchase, however, is likely to require banking or credit schemes. Banking schemes are relatively straightforward once there are herder associations that can deal directly with the commercial banking sector, but credit schemes require more planning.

Livestock credit schemes have few precedents but offer interesting possibilities to facilitate drought offtake and restocking. One possibility, if redeemable credit is extended when animals are bought during drought, is to express that credit in terms of livestock units, either to cover the higher price of animals bought after the drought or weighted in favor of one species or another. Setting up a livestock credit scheme appropriate to the country or area concerned would require a feasibility study. Terms of reference and some of the options that may be worth examining are suggested in Annex F.

Restocking destitute pastoralists is normally best organized through societal restocking procedures. The best value for money (restoring the maximum number of families for the money available) is obtained through restocking with sheep and goats rather than with cattle or camels. Adding a donkey helps to transport goods and water and generally facilitates mobility. Details would need to be discussed with the society that is to be assisted concerning: (a) the number of families to be restocked, (b) the level of that restocking, (c) the capacity of the societal system to cope, and (d) the cost of project support, including monitoring.

If the proportion of families needing assistance exceeds about 20 percent following an average drought, then probably another strategy is indicated (note the section on sustainability in Chapter 6). Otherwise, environmental issues arise only if the basis of restocking is importation rather than redistribution of animals. In principle, societies should decide sources, recipients, and allocations per family, although a project is entitled to set standards and intervene to ensure equitable distribution of benefits. The experience of NGOs such as OXFAM and Heifer International can be brought to bear, although always ensuring that schemes are adapted to the specific society and that the breeds used are adapted to the environment and pastoral system.

10
Phasing of Technical Inputs

Technical intervention in pastoral systems has a long history. Early interventions focused on increasing water supplies, improving veterinary services, formalizing livestock marketing, and encouraging rotational grazing. Sometimes technical intervention went further, to include rangeland improvement, feeding during drought, or upgrading livestock.

When assessed individually, many of these inputs met expectations and many were also popular. Inputs to restrict grazing patterns or livestock numbers were not popular, but other interventions were. Pastoralists appreciated life-saving veterinary inputs and were especially appreciative of extra water. Indeed, borehole technology led to a misconception that deep drilling can cause water to spout anywhere, and so transform the local economy.

But while technology worked, it did not succeed. Development was seldom achieved because technology was applied out of context, with too little attention to how pastoral systems operate and how the inputs themselves interact. This chapter focuses on how inputs interact and on how technological interventions can best be brought on-stream. Relevant technologies are reviewed, and the rest of the chapter then considers how to phase or sequence inputs.

Relevant Technologies

If the development objective is as stated in Chapter 6, then inputs must contribute socially and/or economically while also being environmentally benign and manageable. In addition to conveying direct benefits in the areas just listed, technology can be used to generate indirect benefits, such as when boreholes are introduced into a project earlier than might be necessary in order to gain approval for institutional innovation that otherwise would be withheld.

The relevance of technologies to NRM is summarized in Table 10-1, and each intervention listed is reviewed briefly in the following paragraphs.

Water development is irresistible in arid pastoral areas as a means of easing the hardship of pastoral life, but it also carries environmental risks. Although arguments abound that additional water points enable better utilization of grazing and that any new supply will be put to its appropriate seasonal use, this seldom happens in practice. New facilities that offer a (near-) permanent water supply readily attract permanent users, with sedentarization and loss of mobility, while also disrupting the established equilibrium between wet- and dry-season grazing. 'Better' utilization of grazing causes more grass to be consumed as it emerges and before it has had much chance to photosynthesize, which restricts further growth and leaves little as reserve grazing. Such impacts usually reflect population pressure, but also arise from insensitive design, especially surface water storages.

Surface water storages are unreliable in arid areas because of uncertain rainfall and runoff on which they depend. The engineering response is to build relatively large storages — dams or ponds, according to the site — which then attract settlement during wetter years and prove unreliable during drier periods. Smaller storages, especially covered cisterns that do not induce permanent settlement but which are strategically useful to maintain mobility, are frequently to be preferred. Sites can be chosen where water harvesting can be used to fill such storages.

Water spreading applies only where landscape and streamflow are conducive to diverting water by simple structures onto land suitable for growing grass or crops. This is not a technology to introduce precipitously, but it is valuable under the right conditions to increase food security.

Groundwater development is potentially destructive because of the permanence of the water it makes available. There is usually no harm, and much potential good, in contributing to the maintenance of hand-dug wells. Ecological destruction is usually associated with boreholes, and mostly with single boreholes. To discover and develop a new aquifer that has the capacity to support multiple boreholes — and hence a new dry-season grazing area — can be regarded as good news. But single boreholes, even if conceived as serving a drought reserve, end up supporting a settled population in a state of increasing degradation as more and more people become dependent on one unreliable pump and a falling water table.

Range management has been downgraded on the development agenda over the last decade as the merits of customary pastoral practices became more widely appreciated. Yet many pastoral situations need a range management input to help reintroduce the environmentally sound management practices that pastoralists used to apply, but which are now being relinquished in favor of a more sedentary lifestyle. It is a strange paradox that pastoralists are now increasingly opting for the styles of resource use that critics of past development efforts saw as representing Western bias, and range science is left to champion the merits of pastoral practice. Opportunities lie in several areas:

Grazing management in the arid zone needs to concentrate on maintaining seasonal movement patterns. Where movement is between dry- and wet-season grazing areas, the latter need to be substantially larger than the former. Although

Table 10-1. Technological Intervention in Support of Pastoral NRM

Area of intervention	*Relevance in arid zone NRM*
Water development	*Potentially disruptive*
Shallow well improvement	Usually a top priority
Water spreading	A priority locally
Surface water storage	Valuable if strategic
Borehole development	Can be highly disruptive
Range management	*Benign provided no fencing*
Browse and woodland uses (gums, charcoal, etc.)	Usually a top priority
Fire use and control	A priority locally
Seeding	Locally relevant
Grazing management ('rotational' systems)	Relevant for existing and potential perennial plant associations
Animal health and production	*Neutral if offtake assured*
Veterinary inputs	A top priority
Animal husbandry (hygiene, calf rearing, etc.)	Often useful
Breed improvement	Usually a low priority
Supplementary feeding	Costly and often disruptive
Other areas	
Other commerce	Usually a top priority, e.g,. goods in/gums and honey out
Livestock marketing	Usually relevant
Cropland improvement	Essential in agropastoralism
Wildlife utilization and (eco)tourism	Locally relevant

dry-season grazing areas often appear devastated by the end of the grazing season, they have the benefit of a rest each growing season. This rest can enable perennial grasses to persist where otherwise they would give way to annuals. It is helpful if there is good browse in the dry-season grazing area and if the community has access to a higher rainfall area to use during protracted drought.

The scope for more sophisticated grazing systems depends on conditions — site potential, vegetation type, livestock species involved, and the competition that may exist among pastoral groups for the grazing on offer. Rotational grazing systems work best in semi-arid and wetter areas with diverse vegetation available for manipulation. All areas with potential to support perennial grasses, however, are likely to benefit from grazing regimes built on the principles of holistic resource management (Box 10-1), provided that the ecological guidelines attached to the HRM model are interpreted in relation to local ecology.

Fire is seldom used as a management tool in the arid zone, not just because there is little biomass to burn, but because recovery from burning is slow and valuable perennial plants are often killed along with less useful species. On the other hand, periodic burning can be very helpful to control bush in the semi-arid and subhumid zones. In those zones it may be necessary to persuade authorities to ease current burning restrictions, as Coppock (1994) describes for Ethiopia. In the arid zone, however, external assistance is more likely to be needed to make and maintain firebreaks to control unwanted fire.

Browse can be immensely valuable. Inputs are needed only for unaccustomed uses or if existing stocks are to be supplemented by plantations, which can also help provide fuelwood and other products. Planting and overseeding of rangeland is much more practical in semi-sedentary than in mobile systems. There is often scope for collecting naturally occurring gums and resins or other products as a source of supplementary income.

Animal health and production is another field that experiences a paradox similar to range management — in this case because there is now often a demand for 'improved' breeds which, if not controlled, could seriously impair the innate tolerances that characterize indigenous breeds.

Veterinary inputs are always in demand. The issue in project design is usually the delivery system: how best to devolve health care to communities, involve the private sector, and ensure government backstopping. Building on local interest in health care and medicines may be the best way to encourage herder organizations toward broader involvement in NRM.

Breed improvement may also be in demand, but is unlikely to feature in the early stages of project design. Projects in arid zones are more likely to be concerned with maintaining the purity of indigenous breeds. There may be scope, however, to encourage use of desert-tolerant species and breeds, and genetic improvement within existing breeds using selection through an open nucleus breeding system (Box 10-2).

Animal husbandry is of increasing importance as herders become more dependent on fewer animals than they formerly controlled. Adoption of unaccustomed methods of husbandry, especially in calf rearing and housing and caring for small stock, may be required. Inputs in agropastoral systems also need to focus on the performance of oxen and other draught animals.

Supplementary feeding may need to feature in new methods of husbandry, especially in agropastoral systems, but only selectively and not so as to create dependency on ever-increasing quantities of imported feed. Feeding cottonseed to lactating animals has been shown to reduce calf mortality in agropastoral systems of Mali and Nigeria (ILCA 1994).

Commerce and marketing systems need attention to facilitate trade in livestock and other commodities, including timely destocking and restocking in the context of drought management. The greater initial input is likely to be institutional rather than physical, but there are exceptions:

Livestock marketing facilities may be required in the form of trek routes and associated infrastructure. The extent of investment in

Box 10-1. The Principles of Holistic Resource Management

HRM is grounded on the thesis that sustainable resource management is possible only if all interacting ecological, economic, and social factors are taken into accout in the management process (the 'whole' referred to by Savory 1988). HRM starts with a process of discussion and negotiation during which resource users set a common goal made up of three interacting parts: (a) their desired quality of life, (b) the production to attain the desired quality of life, and (c) a vision of the landscape and ecosystem that will sustain the necessary production.

HRM methods are based on the recognition of: (a) the difference between 'brittle' and 'non-brittle' environments, (b) the importance of animal impact on soils and vegetation, and (c) the importance of the time during which the rangeland is exposed to grazing. The testing guidelines help users to select the proper tools that are expected to improve the four blocks of the ecosystem. HRM is a continuous monitoring-control and replanification process. Although HRM incorporates a powerful planning tool, the setting of a 'common goal' is far from easy with large pastoral groups with extensive and overlapping grazing orbits (authors).

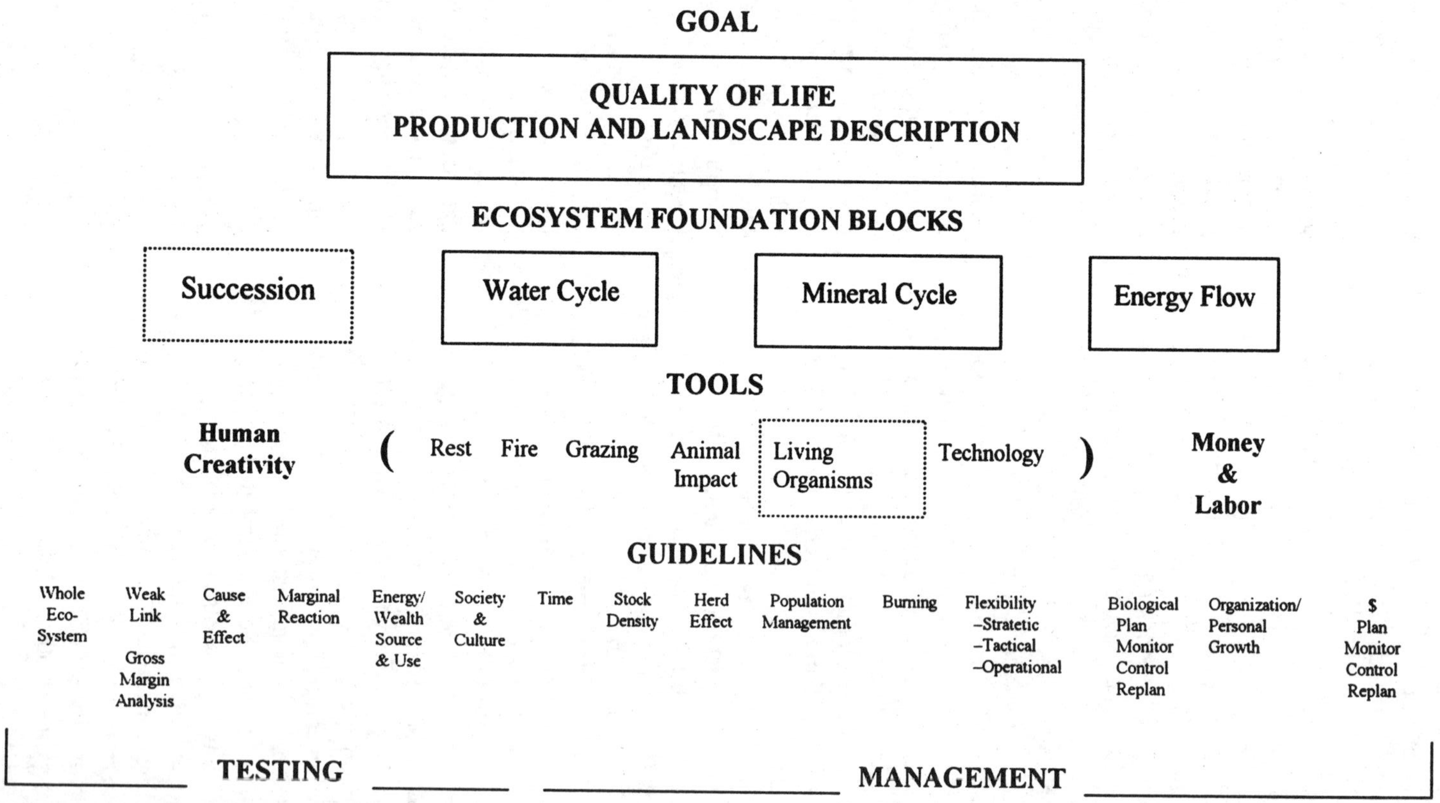

Source: Savory 1988 (model) and Lühl 1992 (text).

infrastructure depends on the strength of private sector interest — sale facilities might be contributed from that side, with watering points and holding grounds along primary routes installed with public funds. Pricing policy and incentives for private sector participation need to be sorted out before investing in infrastructure.

Also important are marketing policies and infrastructure designed to contribute strategically to offtake. Strategic solutions must consider (a) the trade orientation of the pastoral systems served; (b) the categories of livestock that each system has to offer (as opposed to those that need to be retained within the system); (c) the need to ensure pre-drought and crisis offtake; and (d) the distribution of benefits within pastoral society. Urban demand may be strongest for beef, but even if a pastoral system has cattle, it may only be able to offer sheep and goats if it is to remain viable. A pastoral system may derive greater advantage from mobile abattoirs processing half-starved camels and goats into meat-and-bone meal or pet food than from a facility designed to deliver fat stock to market.

Retail stores are often much appreciated by pastoralists in outlying areas as a supply point for domestic goods and sometimes for storing products on their way to market. Small amounts of start-up money — probably to end up in a locally controlled revolving fund — may therefore be a sound investment and more than incidental to NRM if the outgoing products include items such as gums, resins, and honey. Medicines are usually held apart from other items so that they are more secure and can be dispensed by someone with some relevant knowledge.

Wildlife utilization applies only where there is a wildlife resource available for exploitation. Each case must then be examined on its merits. The concept is not new — even when group ownership of pastoral land was first introduced in Kenya in the 1960s, it was argued that the prospect of groups earning revenue from wildlife could be used to encourage wildlife conservation. It is, however, often a long process to change existing policies and overcome entrenched departmental viewpoints. Investment is most likely to arise where pastoralists wish to encourage ecotourism in their area, requiring roads and safari-type accommodation (Annex G).

Cropland improvement is central to development in agropastoral systems, and it is also in and around cropland that most opportunities arise for planting fodder shrubs and seeding grasses. Fodder banks, based on legumes established with low-input techniques on old corral sites and fallow lands, have been introduced successfully in Nigeria (ILCA 1994). Likewise, fodder shrubs such as *Atriplex* and *Acacia* are being introduced as windbreaks in Middle Eastern projects.

Inputs to crop production itself are likely to focus on water-efficient crops and tillage systems, and improved delivery of essential services. Rarely would fodder replace food crops, but leafy sorghum or millet with robust stalks, and possibly cowpea, certainly have a role. Storage systems also need attention so that part of the extra feed produced is available at the end of the dry season for the oxen, donkeys, or camels that are used for land preparation.

Although agropastoral systems offer wide scope for technical intervention, it is necessary to know the system before intervening. Introduction of improved fodder production is frequently impaired by lack of labor at planting time. Perennial fodders may be acceptable under these circumstances when annuals are not. Other constraints also need to be understood. For example, if manure is used as fuel in a region with cold winters (as in eastern Anatolia), that use will take precedence over all others, and a subsistence herd must be defined in terms of the number of animals required to produce a year's supply of fuel.

Sequencing of Inputs

The preparation of likely development paths as part of project planning (Chapter 6) helps to show potential interactions among inputs, such as when one type of input must logically precede another. This logic is critical — it makes no sense to start with inputs that have highly demanding management requirements or with those that risk damaging the environment or the integrity of the pastoral system.

These considerations are the main determinant of the three phases set out below. They might conform with successive two- to three-year stages in a process project, although the speed with which a project unfolds and the precise order in which inputs are made will vary from case to case. Apart from the technical inputs listed, establishing herder organizations and an adequate support system will always be part of the opening phase of project activities

Box 10-2. Open Nucleus Breeding System (ONBS)

The Open Nucleus Breeding System (ONBS) developed from cooperative breed improvement in Australia and New Zealand. By concentrating selection in part of the livestock population (the nucleus), ONBS offers the potential for higher rates of genetic progress than would be obtained by classical within-herd selection (closed nucleus breeding system).

Herd and flock sizes are often small in developing countries, and the basic infrastructure for recording performance and pedigree does not exist, so within-herd selection is not an option. The particular advantage of ONBS in this case is selection concentration so that the performance of a smaller number of animals is recorded, and more detailed information can be obtained on each animal. At the same time, genetic gains achieved in the nucleus are still transmitted to the whole population

The principles of the system are shown below. The nucleus is established by selecting the best males and females from the base herds. Depending on the efficiency of this initial selection, a substantial immediate gain in performance can be obtained in the nucleus relative to the base. Once the nucleus is established, all males selected to be used as sires in the nucleus are bred exclusively in the nucleus. However, a proportion of the females needed for breeding in the nucleus is taken from the base herds. Males and females that are surplus to requirements for breeding in the nucleus are transferred to the base herds. Thus the genetic improvement made in the nucleus is continuously transmitted to the base population.

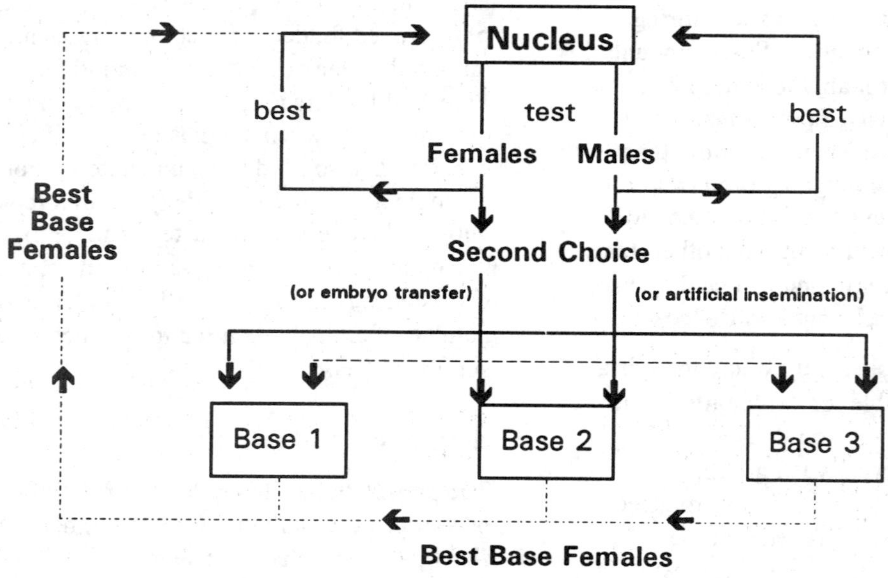

Note: In closed nucleus breeding systems the flow of genes is unidirectional (from the upper to the lower tiers), while in open nucleus breeding systems the flow of genes is bidirectional (from the upper to the lower tiers, and vice versa).

Source: Barker 1992

(perhaps even a pre-project phase). The establishment of a monitoring system will also feature in the opening phase, but that aspect is covered in the next chapter.

First phase inputs are likely to be restricted to strategic water development, animal health care, and possibly livestock marketing.

Strategic water development usually implies restricting inputs in order to maintain the integrity of existing wet-season and dry-season grazing areas (or summer and winter grazing areas in temperate regions). This may involve enhancing the dry-season area with a few new wells or an extra borehole, but more typically it means improving existing wells in that area and improving seasonal supplies in the wet-season area. The aim in the latter area would be to allow a wider distribution of livestock for a longer period before the stock retreat to the dry-season grazing area. Rainfed cisterns may be useful along the way to slow the retreat. Any necessary hydrogeological survey would be undertaken in this first phase, along with the selection of sites for future water spreading.

Animal health care needs to be included in phase one activities because of its importance to pastoral economies and as a point of entry for participatory development. If no new water supplies are included in phase one, animal health care would be the first new venture to offer for community management. The usual strategy is to train herders in basic health care, ensure a supply of drugs under cost-recovery arrangements, and support government veterinary staff in providing necessary back-up services.

Livestock marketing would almost certainly attract support, but with attention to institutional arrangements rather than to physical facilities. Any investment component in phase one is likely to focus on strategic marketing in support of drought management. Commercial marketing would feature as a first phase input only if the pastoral system being supported is already trade oriented.

Other inputs in the first phase would almost invariably include: (a) attention to the rest of the institutional framework; (b) literacy programs and other basic social services; (c) assurance of appropriate policy and legal frameworks for pastoral tenure; and (d) assurance of baseline data to support subsequent monitoring. Further physical development, however, is likely to be restricted to the funding for community-based initiatives (as in the Kenya Arid Lands Project, Box 9-1). In that way, some of the inputs listed next might be brought forward into the first phase of operations.

Localized or second phase inputs focus on specific resources and opportunities for managing resources for food security.

Cropland improvement would certainly feature alongside the components already described in projects addressing agropastoral systems. The expected focus of activities was indicated when describing relevant technologies.

Water spreading and water harvesting are attractive propositions if suitable sites exist, but are unlikely to be pursued among phase one activities unless the community is already experienced in this field. In the absence of that experience, it is best to start with mini-schemes. Harvesting schemes are easier to design and manage than spreading schemes. The former can include harvesting for cistern storage, and in semi-sedentary systems, harvesting for crop production. Harvesting or spreading for grass production in larger schemes might follow later.

Animal husbandry is likely to receive attention once health care is in hand. The focus would need to be guided by analysis of the prevailing livestock production systems, but calf rearing, penning arrangements, and the post-drought nutrition of breeding animals should all repay attention. Most of the inputs are made at the household level.

Woody plant resource utilization often offers better prospects than seeking to improve the utilization of grazing. Even if browse seems fully utilized, added value might come from reserving access or collecting and storing leaves and pods for more strategic use. In addition, stocks of the most useful components of the natural vegetation might be increased, perhaps with the aid of runoff diverted from other sites. Apart from helping communities to optimize browse resources, assistance might be extended to assist in the collection and marketing of gums and resins or medicinal plants. Options usually increase with rainfall and biodiversity, along with scope for using fire to control noxious species. Where woody vegetation is sparse, tree planting and alternative fuel technologies may be needed.

Later inputs, for introduction when community-based organizations are well-established, address the broader and more complex aspects of NRM. The two considered here concern range management. As noted earlier, breed improvement might also be considered along with further expansion of water supplies.

Grazing management is a critical but complicated area in which to intervene. Customary grazing systems usually represent complex adaptation to social custom and territoriality — including responses to factors such as lack of water, seasonality, and consideration for the well-being of livestock (Annex A). To introduce additional ecological or economic criteria can be a trial for both planner and pastoralist. It is easier to intervene locally through a village committee than on a broad scale through one or more associations encompassing greater variation in the factors just noted. Also, although it is possible in wetter zones to apply rotational grazing, there is probably nothing better for the driest zones than seasonal movement between extensive wet-season grazing and more restricted dry-season grazing areas.

Rangeland improvement through seeding or other means may find a place locally, but only if the community is able to rest and protect the treated area from grazing. As always, prospects improve with rainfall, as do prospects for cropping, and in that case there may be more to be gained by investing in cropland improvement than range improvement. A useful review on these issues has been contributed by Bayer and Waters-Bayer (1994). There is also quite a literature on reseeding and overseeding in drier zones (for example, Pratt and Gwynne 1977), although the success rate is rather limited.

11

Process Monitoring

All projects need a monitoring system that provides a permanent record of inputs and outputs. This is a matter of public accountability. A process project, however, also requires a flow of information that enables the project to be adjusted as it proceeds, and this is the principal concern of this chapter. There are two reasons to emphasize process monitoring:

- it should constitute a project component in its own right and requires as much design attention as any other component; and
- its form is not standardized like input-output monitoring, so is not well documented. There is, however, extensive written material on management information systems and on how to plan inquiries in each of the disciplines concerned.

The indicators that are monitored need to be related to project objectives, while also having the capacity to highlight unexpected changes.

Attributes for Monitoring

Since NRM is a product of three interacting systems (Chapter 1), monitoring should be viewed in relation to these same systems. Figure 11-1 lists the main attributes to monitor according to the system to which they are most closely associated. Because NRM is always the product of the interaction between these systems, the column in which an attribute is listed is somewhat arbitrary. In each case, effects must be tracked through the entire matrix in the figure. Thus, although territoriality is listed as an attribute of the user system, it is also a product of the geopolitical system. Similarly, herder organizations are treated as a response to the geopolitical system rather than as a product of the user system. Livestock production is listed as an extension of the user system as though unaffected by external market forces.

Although the following sections offer guidance on the types of data to collect under each heading, the final decision on the attributes and criteria to cover should be made during participatory planning and reflect the objectives set for the project and the perspectives of all stakeholders.

Ecological Monitoring

Two general points that affect ecological monitoring in particular need to be emphasized.

First, all monitoring systems need to be designed so that they focus on what is relevant and avoid being swamped by background noise. In arid zones more than elsewhere, monitoring is liable to pick up an embarrassment of change in floristics and groundcover because vegetation responds spontaneously to a normal range of variation and periodicity in rainfall. If standard criteria and indicators of rangeland health and land quality are applied along the lines indicated in Annex H, significant ecological trends would be lost entirely in a plethora of signals of impending doom or remarkable recovery. The guidelines presented in Chapter 3 should also be considered during operational interpretation of the proposals incorporated below.

The second point is that all monitoring must be organized within a framework appropriate to the subject being monitored. Ecological monitoring needs to be organized within the framework of ecological land units. These should separate major variations in climate and landscape so that each locality where data are collected can be placed in an ecological category.

Figure 11-1. Attributes to Monitor in NRM Projects

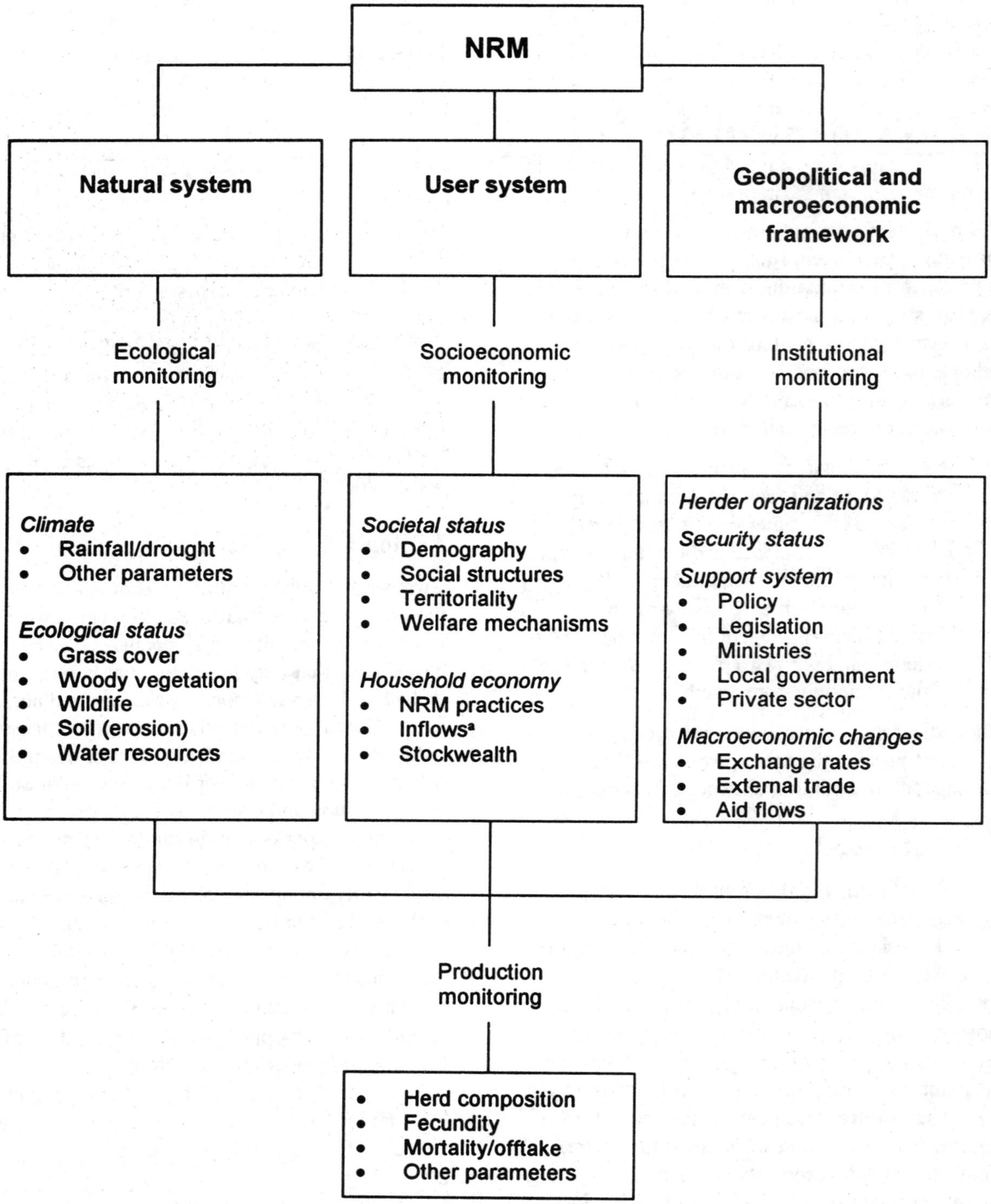

a. 'Inflows' cover cash from employment, remittances coming into the household, support through societal mechanisms, and famine relief.

Key resource sites would be separated at this stage for particular attention. Much of the required data, however, will be forthcoming from drought management (Chapter 9), which has requirements of its own for information on several key ecological and socioeconomic parameters. These data will emerge from that source at more frequent intervals than is required for purposes of monitoring longer-term trends.

Climate is monitored through standard parameters and instruments. Rainfall is the most critical parameter, and projects should expect to add to the existing network of rain gauges. There needs to be one standard rain gauge per 100 to 1,000 square kilometers, depending on zone topography, and the availability of conscientious residents willing to do the recording. Automatic rain gauges and weather stations are not recommended because they are relatively expensive and prone to sabotage. The objective is merely to have a modestly reliable record of rainfall over the project area to which other data and observations can be related. Other climatic parameters are far less critical. There can be situations where excesses of cold, heat, or wind are important, but in most situations the data from the nearest meteorological recording station will suffice. Seasonal fog or mist presents a special problem because it is ecologically vital but tricky to measure. Where fire is being used or controlled, data on wind and air humidity are needed.

Ecological status is easier to conceive than to measure. It is best to record individual features as quantitative justification for any subjective synthesis, rather than attempt a single composite interpretation of ecological status. Satellite imagery does permit an overall interpretation, but only by differentiating areas according to the reflectance of their land surface as determined by the density and greenness of the vegetation cover and the smoothness and color of the soil surface. Neither that interpretation nor an assessment on more detailed criteria has much meaning without being able to refer to records of past rainfall and usage.

Grass cover, woody vegetation, and soil are usually monitored together by one or more recorders who revisit selected spots or transects every few years. More frequent recording is appropriate initially while still assessing the baseline situation, but recording every three to five years is sufficient to assess ecological trends. Close attention should be given to how woody vegetation is monitored. While the presence or absence of species (annual-perennial, noxious-palatable) suffices for monitoring ground vegetation, other indicators must be added for trees and shrubs (density, vigor, and extent of cutting or browsing).

Wildlife are normally assessed independently. Specialists are needed to relate habitat status to wildlife scenarios. There are many scenarios that may be relevant, and wildlife monitoring should not be confined to areas where there is interest in conserving or restoring the original fauna. Ensuring habitat suitable for lesser fauna, or maintaining plant species of special economic, genetic, or esthetic interests, can be legitimate objectives. As with vegetation monitoring, details should be formulated on site. Annual or biennial monitoring may be needed in order to track population dynamics.

Water and feed resources are routinely monitored to give early warning of drought. These data supplement rainfall records to build a picture of drought cycles useful to adjust project strategy. For purposes of determining ecological trends, however, the most significant hydrological indicators are spring flow and the discharge and rest level of wells and boreholes. While wells can be monitored by their supervisors, borehole testing requires periodic input by a hydrogeologist to interpret recharge-discharge data and assess what is happening to the regional aquifer.

Socioeconomic Monitoring

Although no branch of monitoring is intrinsically more important than another, socioeconomic monitoring is operationally the centerpiece of process monitoring. If project objectives are similar to those adopted in this document (Chapter 6), then changes in the circumstances of the user system — whether attributable to the project or not — could profoundly affect the direction in which that project is guided.

Socioeconomic monitoring needs to be undertaken within two frameworks: that represented by pastoral society as a whole, and that established by differentiating households of different wealth and social status (Figures 11-1 and 11-2).

Societal status is a more vague concept than ecological status, and its attributes cannot all be quantified. Nonetheless, assessments need to be made periodically and at about the same frequency, unless there is need to step up monitoring because sudden disruption is threatened, such as loss of land to outsiders. A demographer or social anthropologist and the community itself should also be involved in the monitoring process.

Demography is an important area for monitoring in order to keep track of population growth and emigration, and to update population trends and their implications for the future. If a demographic study is part of project planning, then the preparation of guidelines for future monitoring can be required of that study; otherwise the local university should be able to advise. One aspect that is difficult to judge from afar is the extent to which ethnicity should be incorporated in the monitoring framework.

Figure 11-2. Interactions for Monitoring at the Household Level

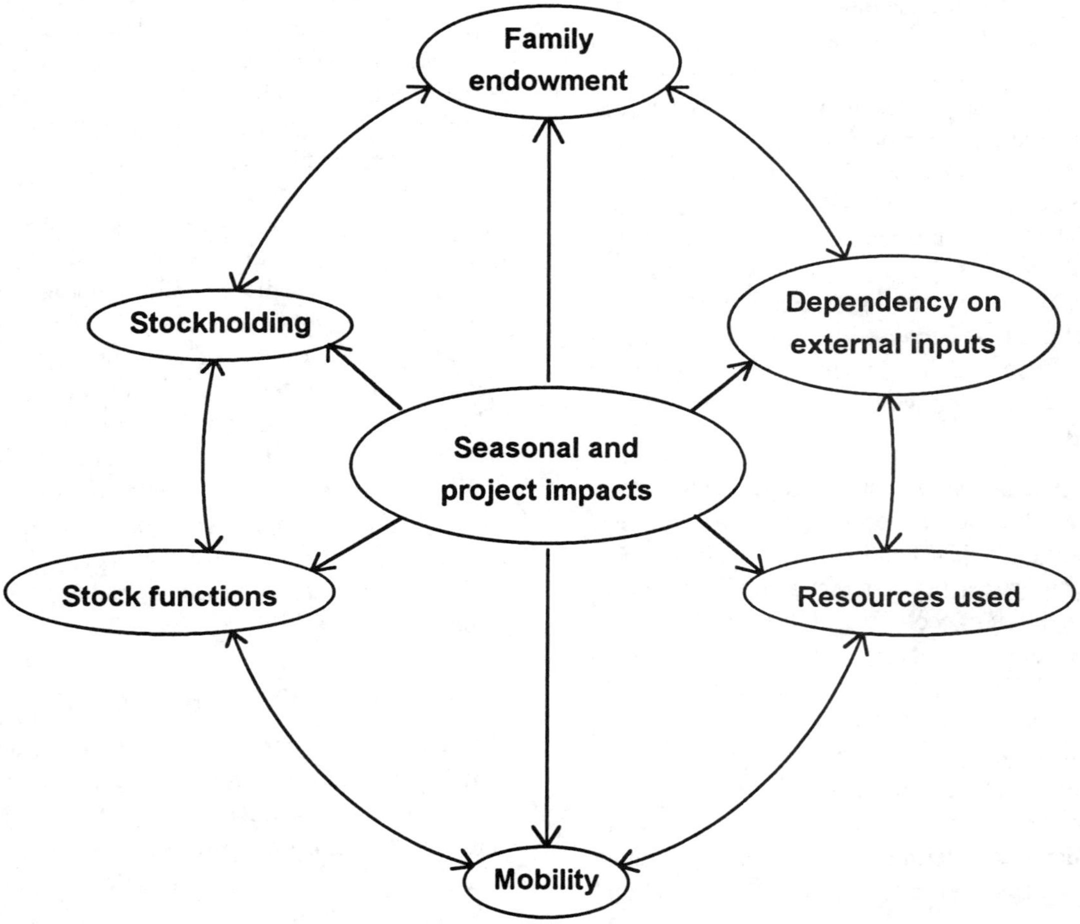

Sometimes this is a non issue or is glaringly obvious, but there can also be nuances of clan or ethnic relationship that are fundamental to the functioning of society.

Social structures and territoriality are likely to be monitored through dealings with herder organizations (covered here under institutional monitoring). If these organizations are not built upon customary social structures, separate monitoring of the latter will be needed. A loss of effectiveness in customary institutions (as measured by breakdown in customary controls and conflict resolution) could have wide implications and call for a shift in project strategy. Even more damaging would be loss of territory or mobility. These are fields of monitoring in which project management might wish to participate directly. Key parameters are: (a) settlement or privatization of resources against societal interests and wishes, (b) the marginalization of elders in decisionmaking processes, and (c) decline in the effectiveness of resource tenure and welfare support mechanisms.

Welfare mechanisms need to be monitored as and when they are brought into intensive use. Even if constantly in operation, the most revealing time for assessing their effectiveness is when they are under increased pressure. Routine monitoring of a project's drought management component may suffice, although welfare mechanisms serve not just drought victims (Annex A). Criteria for deciding who should get how much of what form of assistance vary with pastoral system, demand, and availability of famine relief.

Household economy needs to be assessed within a framework established through wealth ranking. Sometimes social status unrelated to wealth must be incorporated in the ranking. Since each stratum differs somewhat in its NRM practices and is influenced differently by project inputs, household-level monitoring is essential to the operation of a process project. Since household-level monitoring is also labor intensive, the amount of information collected per household and the number of households included in the sample are critical decisions. The size of the sample must be decided locally, but it is possible to suggest ways of reducing the amount of data collected.

NRM practices can be the focus of one monitoring package. However, the monitoring should differentiate the practices of: (a) poor households with few or no livestock, (b) households with a subsistence livestock holding, and (c) wealthy households with large herds or flocks. Group discussion involving all strata is useful for exploring interdependencies and differing perspectives, but the main data should come from enquiries with individual families in each stratum. It is not advisable to pursue detailed enquiries with more than ten families per stratum, partly to contain costs, but mainly because it is difficult to build and maintain personal relations with more than about thirty families in total. The aim is to establish how families of different size, composition, and resource endowment vary in their dependence on NRM, access to resources, mobility, and drought strategies. The matrix of these interactions includes the factors just indicated plus the effect of season and project inputs (Figure 11-2).

Initially, enquiries would need to be undertaken once or twice a year, but the frequency can be relaxed once a picture of the normal range of variation has been established. However, a capacity needs to be retained for returning to assess the impact of any extreme event that occurs — particularly drought, but also loss of land or a potentially influential intervention.

Inflows into households need special attention in order to establish what can be done to help families become more independent of external assistance, and to assess the impact of initial moves in that direction. All inflows are relevant, including remittances and famine relief, although particular attention should be given to inflows from paid employment and societal support systems to provide information to guide new initiatives in those areas.

Other parameters would need to be monitored if trade and cash flow are relevant to project objectives, or if inputs are made to cropland in agropastoral systems. It is wise to be selective, however, since the more data that are collected, the greater the likelihood of survey fatigue and the greater the burden of data analysis.

Production monitoring is treated here as an extension of socioeconomic monitoring because it is only really informative when organized at the household level within the framework already described. The methods, however, are quite different from those used in other types of monitoring, and require an input from an animal scientist familiar with systems studies. The principal parameters include:

- herd and flock composition, particularly reproductive capacity;
- fecundity, reproductive disorders, and mortality as affected by breed type, age, season, and management;
- offtake as determined by seasonal conditions (also establishing if low sales are due to market deficiencies or overriding benefits of not selling); and
- other parameters relating to animal husbandry and animal health care, as relevant to the project being monitored.

Institutional Monitoring

Monitoring the functioning of institutions and organizations involved in project activities is especially critical during the first phase of operations. Close and direct scrutiny at that stage might be relaxed later to the annual review of records kept by each organization. An essential step, however, is to draw up a list of the functions of each entity, and to adopt indicators appropriate to those functions.

Herder organizations can be assessed ultimately through ecological and socioeconomic monitoring that will show the effectiveness of organizations on the ground. Meanwhile, the institutional status of herder organizations is revealed by:

- membership relative to constituency;
- records of meetings and decisions (frequency, attendance, and relevance of decisions relative to needs and mandate);
- extent to which members are aware of decisions taken and place value on what the organization is doing;
- evidence of complementarity with customary institutions and relevant public and private sector bodies; and
- level of indebtedness and related controls, considered by object of expenditure and possible sources of income.

A specific role of monitoring is to identify where performance is constrained by the terms of reference or legal constitution of an organization, and where the constraint lies in training or counseling. A project that itself is institutionally effective should then be able to make or broker the appropriate adjustment. At the same time, allowance should be made for the value systems and procedures of the society. If, for example, organizations do not avail themselves of credit facilities, it may reflect indecision or incompetence, but it may also be that the society does not approve of indebtedness.

Security status is more than an institutional issue, but belongs here because it often defines the institutional environment for pastoral development. When armed conflict disrupts mobility and the flow of development inputs, other institutional issues fade in significance. Projects therefore need to keep security risks under review and may need to consider insecurity when planning to monitor or interpret results. Whether it is possible or worthwhile to actually monitor security status depends on the form that insecurity takes. If there is a known flash point, then it may be possible to obtain advance warning by keeping this under review, even if the monitoring system has to be informal.

Support systems should be monitored in a manner similar to herder organizations, although with allowance for their different functions.

Policy and legislation appear in written form and so are easily monitored. But less easy to monitor is how the *relevance* of policy and law changes as social and economic circumstances change. This is a necessary part of keeping up-to-date on reforms needed to support pastoral development. It may be sufficient to incorporate additional lines of enquiry every few years in the monitoring of herder organizations, but more elaborate monitoring may be needed if a project has responsibility for advising on policy or legal reform. A lawyer needs to be involved if monitoring land litigation or drafting bills is part of the project brief.

Ministries and local government often feature as partners in development, and may have received substantial inputs to improve their institutional capability. Monitoring performance can usually be based on the annual review of operational capability, plus field studies every two to three years of the extent to which the needs of pastoralists are being met. Unless organizations that are being supported under a process project report in timely fashion or submit themselves to scrutiny, their funding would need to cease because there is then no basis to define their continuing role.

Private sector contributions need to be monitored according to the service delivered. Different criteria apply, for example, to banking services, supply of drugs, or inputs to water development. It helps if at the outset there are agreements with each major contributor detailing the necessary records of services delivered. If client confidentiality or other reasons prevent detailed information on the distribution of services from being released, then proportionately more effort will need to go into field monitoring of the effectiveness of private sector inputs.

Management Implications

Managing a monitoring system of the type described involves: (a) programming the system so that all parts come on stream in complementary fashion, (b) coordinating the system in such a way that there is a timely flow of digestible data to relevant parties, and (c) canalizing results into project operations.

Someone must have overall management responsibility. That person should be able to comprehend the project and the monitoring system as a whole, and should be part of the project management team. Even where there is a ready-made monitoring team in a participating agency, there needs to be someone dedicated to ensuring that monitoring meets project requirements.

Programming would seek to involve all relevant organizations and field personnel in order to minimize the project's direct role in data collection. The actual time given to data collection is small when considered in terms of individual components, so there should not be strong resistance to participation. Assuming that three or four principal participants have been identified, these could form the basis of a working group to complete the initial planning, identifying any additional skills that need to be absorbed or contracted. The same group could also establish the cost to the project, working within the total budget established during project preparation. That cost would depend on project circumstances, but it would be reasonable to expend up to 5 percent of the budget on keeping the project on course.

Coordination is needed in two areas:

Data collection should be organized through a time sheet that enables the coordinator to remind each participating group when a task is due. The need for requisitions and reminders can be reduced by ensuring that the working group referred to above continues in existence and meets on a regular basis. Regular discussion is also a means to adjust monitoring as circumstances and experience require, and of airing impressions gained in the field.

Data analysis is likely to cause more problems than data collection. As much analysis as possible should be done as close as possible (in terms of time and personnel) to the point of collection. Nonetheless, the project needs to have the capacity to undertake or redo as much of the data analysis as necessary. Results must be reported promptly according to a schedule established in parallel with the time sheet for data collection.

Canalizing results begins with exchange of data so that relevant project staff and professional advisers receive the results of monitoring in time for personal review prior to formal review. The results of monitoring are then fed into project operations through: (a) annual or eighteen-month process reviews, and (b) analytical workshops held every four to six years for retrospective review of progress (for ten-year projects, these would comprise the mid-term and final review). In addition, full data sets should be left by each project to guide new projects.

The main feedback to the monitoring system itself would come from formal process reviews. Although the main purpose of these reviews is to chart the future course of the project in terms of strategy and investment, they are also likely to specify adjustments for the associated monitoring. This is also the time when additional studies might be conceived to pursue a trend or opportunity identified through the monitoring process.

Conclusion

12

Broader Implications for International Agencies

Application of the guidelines from earlier chapters lead development agencies into procedures and activities that lie outside the normal confines of sectoral projects. This final chapter examines four of the broader aspects of aiding pastoral development:

- management issues associated with the process approach;
- improving baseline knowledge of pastoral systems and their environment;
- assisting technology development; and
- overcoming attitudinal problems about pastoralism.

Managing Process Projects

One feature of the process approach is that a project commitment is made on the basis of an indicative cost, supported by a statement of likely development paths. Detailed costs are available only for the first phase of operations, with the rest determined by monitoring. The result may not be much different from the current practice of adjusting and extending a blueprint project as a result of supervision, but it is a significant step for a bank to fund projects whose dimensions are known to be uncertain.

Given the increasing variety of project arrangements, it may be that this additional variant can be allowed without subjecting it to formal policy review. If it were judged unacceptable, then it is better that the World Bank avoids process situations altogether rather than continue trying to force problematic areas of rural development into a standard project mold.

On the assumption that a process approach will be used, the following paragraphs summarize the main implications for task managers. As will be seen, there is not much that is new, although existing procedures are combined in a different way.

Project preparation will vary in time and inputs as in any other type of project, although it will take longer overall because of the need to categorize the pastoral systems and define the scope of the project accordingly. More time than average is needed for (a) retrieving and analyzing existing data, (b) conducting PRA, and (c) preparing likely development paths.

The minimum period from identification to pre-appraisal is about one year, with another six months for appraisal. It is better to involve people familiar with pastoralism than to try to incorporate representatives from each of the ten or so relevant specializations. On that basis, an allocation of twelve staff months should suffice, inclusive of consultant time but not counting official and community representation.

A pre-project facility is usually essential to fund extra field studies and sometimes other inputs once pre-appraisal has shown that a basis for a project exists. Other inputs are difficult to predict, but studies are needed to assess the capacity for cost-sharing and other strategic aspects, as noted in Chapter 8. About eight months of staff or consultant time should be on call, with more accessible if needed. Studies alone, without capital input, may not need a project-tied facility (see Portfolio Management, next page), but the principle of pre-project funding, from whichever source, certainly needs to be assured.

Project implementation proceeds within a time frame and budget ceiling determined at appraisal and negotiations. A process approach does not mean that projects are necessarily longer term than standard projects — usually they would be, but the provision for regular review of 'what happens next' actually provides more opportunities for terminating a sterile project than normally would arise.

Funding arrangements would probably be left open for decision during negotiations. There are several reasons for not placing pastoral development projects in a preconceived financial mold:

Institutional arrangements could follow one of several models (Chapter 5), ranging from one executing agency to an interdepartmental consortium. Local government and/or NGOs are likely to be involved, with provision over time for devolving financial and management responsibility to herder organizations.

Capital content could be relatively small, with a low economic rate of return. Apart from facilities for executing agencies, capital items could be restricted to limited road improvements, groundwater development, and service centers, but even these may come on-stream mid-term in the project rather than at the outset. More critical to project success will be the input to institutional development, studies, and monitoring. Levels of cost-sharing and subsidy need to be decided for each project or pastoral system according to its situation.

Cash flow is episodic — the outlay can be predicted as accurately for the first phase as for any project, but thereafter release of funds would be conditional upon agreements reached at regular project reviews.

Review meetings need to be held every year or two. The frequency cited in earlier chapters is twelve to eighteen months. The object is to plan and approve the forward program on the basis of progress, heavily influenced by data from the project monitoring system. The actual frequency of meetings and the composition of the review group are important matters for task managers to keep under review. Although the monitoring system (described in detail in Chapter 11) is expected to provide all the information needed for decisionmaking, supervisors should assess the monitoring system itself to ensure that it is focused on essentials and is cost-effective.

Portfolio management becomes increasingly important as the variety and style of individual projects becomes more diversified. It has already been suggested (Chapter 4) that range-based projects need to be categorized according to whether their aim is environmental conservation, livestock production, pastoral development, or emergency relief. The categorization is important because requirements are different at preparation and during implementation (Table 4-1).

With such extreme differences, a lot of backtracking is involved if a project is wrongly categorized at the outset. It is usually clear to which category an intervention belongs, although intermediate situations do arise:

Livestock projects are usually associated with semi-sedentary systems under semi-arid to subhumid conditions. The same conditions, however, could call for a process approach if the system is not already commercialized — or even an emergency operation if population pressure is exacerbated by drought. The possibility of subsidy is entered in Table 4-1 on the assumption that elements of rural infrastructure (roads and service costs) are not charged to livestock production. An economic rate of return (ERR) of around 30 percent might apply to a project where inputs allow the marketing of livestock that otherwise would not be marketed. Some livestock projects have claimed an ERR of over 50 percent at appraisal, extending to infinity in the case of Kenya Animal Health Services, but performance has never reached these heights.

Pastoral development projects always require a process approach, although quite often components could have economic justification, such as a marketing component or, under some conditions, drought management and technical services. The cost of the monitoring that accompanies a process project, however, should be included in any calculation of ERR; the monitoring is not a separate or optional input.

Emergency projects are not in the mainstream of World Bank lending, but are not excluded either. Morocco, Kenya, and Zimbabwe have recently had emergency drought recovery projects. Justification is greatest where the emergency operation is followed by a longer-term input to pastoral development, as in the case of Kenya. A short-term project of the type cited in Table 4-1 gains nothing from a process approach unless it is a prelude to a more substantial project to follow. A longer-term emergency project aimed at the relocation of pastoralists would, of course, need a process approach.

It is easier to assimilate a variety of project types if they are conceived as components of a managed country portfolio. The concept of managed country portfolios has the advantage of assuring coherence in World Bank operations and national development. An added advantage for pastoral development projects is that it is easier to encourage policy and legal change in pastoral land tenure if the encouragement can be presented within the context of a total program rather than as an isolated pastoral issue. Finally, a country program framework facilitates the liaison necessary to separate initiatives that are better left to other donors and NGOs and those of a scale to warrant World Bank involvement.

Improved Understanding of Pastoral Systems

Perhaps the most common risk attributed to pastoral development projects is drought, but this is a superficial interpretation. The real risk lies in superficial knowledge of how pastoral systems and ecological processes operate.

Systems studies have been referred to at various points in previous chapters. The term is used to denote enquiries to establish how systems of resource use function and how they respond to change and intervention, including opportunities for diversification. In the context of pastoral systems:

The system comprises the people, their resources, and the environment in which they operate. The environment comprises natural, social, and geopolitical components.

The enquiry is interdisciplinary, conducted by a team that is able to integrate individual skills and work with the community. The enquiry is usually conducted in stages over more than one year, and may culminate in testing interventions.

The orientation is usually towards production, examining how households and larger groupings utilize resources for purposes of livestock or crop-livestock production. This means that the study is a form of farming systems research (FSR), although that term is avoided here because much of FSR is just socioeconomic research with token levels of ecological and biological input.

The result is a report on the dynamics of the system(s) that can guide future research and development. Examples include the ILCA Systems Study series, several of which concern dryland pastoralism (Wilson, de Leeuw, and de Haan 1983; Solomon Bekure *et al.* 1991; Coppock 1994).

More of these studies are required. Apart from their value to the development of the actual systems under study, wider coverage would help both theorists and development planners escape the trap of working to a generalized model of pastoralism. Three or four in-depth studies in each of the categories of systems described in Chapter 2 would counter the overgeneralizations that prevail at present. Studies of desertic pastoral systems and of confined vs free-range systems across an arid to subhumid gradient are particularly needed. More studies will show the individuality of pastoralists' responses (as determined by ethnicity, education, and relations with the State), as well as commonalities linked with climate and resource availability.

Ecological processes need to be better understood for much the same reasons that present generalizations are an impediment to pastoral development. Recognition of disequilibrium is certainly better than an assumption of equilibrium, but better still is to regard both as snapshot interpretations of ecological processes at work over time (Chapter 3).

Even casual observation over a few years is sufficient to show that vegetation and productivity vary seasonally in the arid tropics. Observation over a few decades will then demonstrate that these changes are not random, or at least not without reason. For example:

- Arid zone vegetation can cycle through several forms, with *Commiphora* trees becoming senile and giving way to grassland until the grassland is invaded by *Acacia*, and then by *Commiphora* again.
- Semi-desert grassland can cycle through perennial and annual forms, with perennial grasses fading away during dry decades and returning again during wetter cycles.

These changes occur as one state shifts to another, speeded or slowed by the intercession of a really wet year, a hot fire, or (with *Commiphora*) a surge of elephant activity. It is just not plausible to write plant succession out of arid zone ecology. Savory (1988) is nearer the mark when he invokes a scale of brittleness rather than the equilibrium vs non-equilibrium concept, but even this has its limitations, as indicated in Annex C. As also noted there, it is likely that chaotic behavior within ecological processes is accentuated by aridity.

Research on ecological processes is needed to sort out these ideas and inconsistencies. Placing a few ecologists in the field for a twenty-year spell would help, although computing power is needed to explore chaos and speed the progress through ecological modeling. Such clarification will not necessarily make NRM any easier, but at least it would allow planners to complement the indigenous knowledge of pastoralists.

Institutional responsibility for these types of study is not easily allocated.

Systems studies are in principle suited to national research, although it can be difficult to form and maintain interdisciplinary teams from government sources. Universities offer better prospects if someone else is paying and faculties are communicating, but for comparative analysis of systems across ecological gradients (as recommended above), it is still necessary to look to international research. As an interested party, the World Bank should bring its influence to bear through the Consultative Group on International Agricultural Research.

Research on ecological processes lies upstream of most organizational mandates so that it is less clear who should take the international initiative. Such an initiative is needed, however, since ecological research is not a priority for national systems. University linkages — bridging the North-South divide and combining arid zone interests — are one possibility.

Technology Development

Technology development has bypassed pastoralists. While technology has less to contribute in the arid zone than in zones of higher potential, not much effort has been made to understand or meet pastoralists' needs. Under arid pastoral conditions, technological inputs are best categorized according to whether they (a) add to the supply of stockfeed and water or (b) improve the utilization of what is already available. The principal areas of intervention are listed in Table 12-1 according to these categories, with options discussed below:

Increasing the resource base is not often practical. Additional land is seldom available (outside the tsetse belt of Africa), and there are not many untapped aquifers. The hazards of plowing up more rangeland for cultivation and increasing groundwater extraction further reduce options.

Seeding or planting rangeland has attractions locally, mainly in temperate or semi-arid to subhumid areas and in semi-sedentary pastoral systems. Testing plants for those areas has been extensive, and the main priority for further research is to seek legumes adapted to a wider range of conditions. Undersowing with legumes has the advantage of upgrading soil fertility without heavy dependence on fertilizers, which are usually too costly to apply to rangeland. Research in agroforestry, for fuel as much as for other products, may have higher priority than range improvement *per se*.

Diversifying use of cropland offers great scope in agropastoral systems. The first stage of diversification can usually be supported by off-the-shelf technologies, but there is need for research into more innovative options for later use. The focus needs to be on water-use efficiency and plant material that either combines food products and stockfeed or that will fill identified niches in the farming system. Research on tree crops should not be forgotten.

Mist capture is only an option in areas which experience a misty season. Under those conditions, mist capture offers a rare opportunity to actually increase water resources. Wire screens can collect several liters per square meter per day during the mist season. However, large screens are cumbersome as well as costly, so the method is unlikely to provide more than strategic domestic reserves. Nonetheless research is warranted in appropriate regions on screen designs and harnessing mist collection by trees for their own benefit and that of plants in their vicinity.

Intensifying the use of existing resources is the essence of NRM. There is a wide range of management techniques that theoretically can be brought to bear, but only a few of wide application.

Grazing management research is problematic because it is difficult to simulate pastoral practices experimentally. One possibility to overcome this problem is to use holistic resource management in a semi-experimental mode — applying HRM guidelines thoughtfully as primarily a participatory learning process, mainly in arid to semi-arid areas with perennial grassland. Under very arid conditions, monitoring movements and resource usage is all that is needed initially, while in wetter zones there is more scope for devising grazing rotations to meet specific needs. The latter rotations would be designed around seasonal growth patterns and the palatability of target species, and often the need to allow rest and recovery before and after periodic use of fire.

Burning is not normally a recommended practice in arid areas, although it may still be the

Table 12-1. Strategies for Increasing Availability of Stockfeed and Water

Resource	*Increase supply*	*Intensify use*
Rangeland		
Area	———Usually fixed, unless won by conquest———	
Grazing	Undersow legumes	Use grazing rotation
	[Use fertilizer]	[Burn thickets]
Browse	Plant browse	Manage access; harvest or use 'browse-plus' [a]
Fuelwood	Plant woodlots	Improve fuel use
Cropland		
Area	Plow up more rangeland	Conserve soil/water
Crops	Diversify crops	Conserve fodder
Water		
Groundwater	[New wellfield]	[Increase pumping]
Streamflow	Protect watershed	Water-spreading
Rainfall	n.a.	Water-harvesting
Air humidity	[Mist capture]	n.a.

n.a. Not applicable.

Note: Underlining indicates strategies of particular potential; brackets denote strategies constrained by cost or limited availability of the relevant natural resources.

a. See next page.

most practical means to control thicket formation on run-on sites. If more were known of the tolerance of individual plant species to different intensities of fire, it would assist in the strategic use and control of fire. This is not a plea for more research, but rather for a more precise approach where burning is under investigation, such as recording fire temperature and responses by species and age-class.

Browse utilization is probably the most rewarding area for research. There are two contrasting issues:

- Because heavy utilization kills palatable browse plants without the rapid replacement that occurs with grasses, there is need for more information about how plants (by species and age) react to browsing and cutting intensity.
- Because many woody species have low palatability, there is need for more information about what determines palatability and how palatability and utilization can be improved.

This second problem area is particularly wide. It opens possibilities for managing access to coincide with season of value, collecting and curing browse material (at least from large-leafed or heavily-podded species), and dosing livestock to improve the utilization of browse (along the lines of 'browse-plus,' Chapter 10). Also, the more that is known of naturally-occurring browse, the easier it is to select and introduce additional plants that fit local needs.

Conservation of feed from cropland is a more restricted field for research, but nonetheless useful in order that the storage of crop residues and other products becomes more practical for agropastoralists. The conservation of materials from cropland is usually more practical than making hay from rangeland, although the latter should not be overlooked. The value of conserved material can sometimes be increased by chemical treatment, and always by rationing it by class of animal and season.

Water harvesting probably ranks with browse utilization as the most rewarding area for research. Research on actual harvesting systems need not extend much beyond seeking novel engineering solutions to old problems, but there is also need to ensure that the harvested water is put to best use. Sometimes the water will be stored for drinking, but harvesting runoff for crop or fodder production is also a possibility, requiring selection and testing of appropriate planting material.

Institutional responsibility for technology development is shared between international centers and national programs. The only point to stress here is the need to ensure that the bulk of the research effort is conducted in or in close association with the pastoral systems that are being served. It is easy to say that research focuses on real problems, but less easy to ensure that constraints are being addressed rather than the ambitions of researchers.

One way to ensure requisite research is to write a research component into development projects, or at least to invite active participation by relevant research organizations. This approach has not been examined in detail in the guidelines, but it is one to bear in mind during the preparation of major projects. It sometimes meets with resistance on the grounds that researchers are fully committed to 'real' research, but there is nothing more likely to yield useful results than to focus research on the needs of ongoing development.

Overcoming Attitudinal Problems

'Attitudinal problems' drew comment in Chapter 1 as a major constraint to pastoral development. Systems studies help to correct false images, but it is still necessary to ensure that understanding of pastoral development improves.

Information networks are a step in the right direction, although one feature of networks is that they tend to attract participants with academic interests rather than those with policy and field responsibilities in pastoral development. The most useful are probably the ODI Pastoral Development Network (currently under threat of closure) and some of the country networks. Those with a disciplinary bias are more helpful to career development than to pastoral development.

Selective dissemination of information (SDI) is provided by a number of documentation centers, including the International Livestock Research Institute (ILRI).[1] The standard procedure is to record each participant's interests and then furnish summaries of whatever documentation is relevant to each profile. The problem with this procedure is that accessioning procedures are so generalized that the information provided still contains a high proportion of information of marginal relevance to the recipient. This defeats the purpose, and often what is received remains unread.

Short courses are much more likely to instill an in-depth appreciation of what is involved in pastoral development. Several types of courses are useful, but the essential message should be that it is necessary to:

- appreciate the diversity of pastoral systems,
- make the most of available data and past experience, and
- apply a process approach to new pastoral development.

The rationale for focusing on these themes is self-evident from earlier chapters, but the middle one needs elaboration — since scant use is made of past records and project experience. Even unsuccessful projects can be very informative if the steps that went into the planning and implementation are analyzed in detail. The tendency to write off old projects as attempts to impose western ranching on pastoralists misses the point — even if the concept is changed, many of the barriers to progress remain the same. It is still necessary to ensure that pastoral land is not grabbed by others, and to find the means to delegate management responsibility to

1. The International Livestock Centre for Africa (ILCA) merged with the International Laboratory for Research on Animal Diseases (ILRAD) to form ILRI.

Box 12-1. Incorporating Past Project Experience in Short Courses on Pastoral Development

Full use should be made of the experience in Iran, Kenya, Syria, and Tunisia, all of which have a history of development efforts extending back over thirty years. The Kenyan experience is particularly instructive in that it sought to incorporate new legal instruments to recognize customary rights.

A course of the type proposed would also draw on other examples, but could, for example, include the experience of group registration in Maasailand, with attention to:

- the baseline situation (1966);
- stages in drafting enabling legislation;
- changes incorporated at World Bank appraisal;
- subsequent land adjudication;
- institutional responses to adjudication;
- issues related to group constitutions and counseling;
- impacts on schooling, resource management, etc;
- loan applications, supervision, and repayment;
- environmental, social, and economic responses and impacts;
- the predicament of the second generation;
- stages in the trend to individualization;
- impacts of individualization;
- lessons learned;
- exercises in redesigning the program; and
- inputs now required to help title holders (1996).

Without studying the steps taken to assess ecological potential, social-territorial organization, and stockwealth, as well as to delegate responsibility, the present generation of planners and managers is unlikely to do any better than earlier efforts.

pastoralists. Short courses could usefully dwell in detail on past project experience in these fields (Box 12-1).

Educational systems must also be targeted if attitudes are to be changed. This may lie outside the direct scope of pastoral development projects, but World Bank support for the education sector provides an opportunity to ensure that curricula are attuned to development needs. Information on how aridity affects development needs to enter curricula at all levels of education.

Universities could be the venue for the short courses already described. University staff may also be invited to participate in process monitoring. Through such linkages the World Bank can play a part in encouraging inter-faculty collaboration that is often lacking. One of the most disappointing aspects of university bureaucracy is that faculties so often work in isolation and miss the opportunity to lead interdisciplinary activity and thinking. Indeed, universities must bear responsibility for the dominance of reductionist thinking that is now the bane of managing and developing complex systems. No opportunity should be lost to help universities fulfill their intellectual potential.

School curricula need to include more ecology at an earlier age. By secondary school, curricula must ensure the subjects needed for career development. The primary school years are the time to build on what pastoral children already know about herding. It can do no harm when children start school for them to learn that they are not ignorant, and that people earn PhDs in what the students already know, except that the PhDs call it ethnobotany. To learn something about arid zone ecology and civics based on pastoral organization would not be detrimental to those who leave pastoralism, and could help those who remain to be better pastoralists. The starting point, of course, must be teaching the teachers.

Annexes

A

Customary NRM as Practiced by Pastoralists

How pastoralists practice NRM is conditioned by many factors, some relating to the available resources and others to the needs of the people. This annex focuses on just a few priority features, starting with access to resources, which is a precondition for management and the principal variable that tenure systems seek to control.

Access to Resources

Access assumes particular importance in pastoral situations because:

- pastoral groups overlap and sometimes compete for the same resources;
- areas and resources vary in their seasonal value and accessibility; and
- social differentiation precludes some pastoralists from accessing resources that are available to others.

In addition, these territorial, seasonal, and societal dimensions of access affect NRM in different ways.

Territorial access is normally secured for pastoralists through the numerical or military strength of the larger societal grouping to which they belong. Access beyond that homeland may come through negotiation or infiltration, but the integrity of the homeland itself usually depends on the continuing ability of the group-in-residence to repulse outsiders. Vacuums tend to be filled, unless the State intervenes to uphold the *status quo*.

The higher-order groupings that maintain territorial integrity are not necessarily the groupings that are most influential in the day-to-day exercise of NRM. Pastoral societies are multilayered and well-versed in the practice of subsidiarity so that families and neighborhood groups have considerable freedom in making their own management decisions. At the same time, strands of authority governing spirituality and ethics quite frequently affect NRM. And whoever is making the decisions, customary controls usually apply more rigorously in homelands than in peripheral areas.

Seasonal access is governed by patterns of climate and hydrogeology and associated biological rhythms. Regional patterns preclude pastoralists from using temperate highlands during winter, floodplains while inundated, and waterless areas except after rain when stream flow or standing water are available. Biological rhythms cause pastoralists to avoid seasonal occurrences of biting flies, keep livestock away from predators and other high-risk areas during parturition, and seek periodic access to areas with salt licks or plants of specific therapeutic or nutritional value.

Homelands that have not been curtailed during their formation or subsequently reduced are always of a size or distribution that accommodates such seasonality, at least in most years. If not contiguous, then dry- and wet-season grazing areas, or summer and winter areas in temperate regions, are connected by established routes of transhumance. Customary NRM hinges on seasonal access and is thrown into disarray if that access is disrupted.

Societal access — meaning who in society enjoys access to resources — relates principally to family stockwealth. It is usual for group stockwealth to be concentrated in a few hands, with the majority of households having no more than a subsistence holding. Such imbalances are accentuated by population pressure and drought, and by the growing practice of absentee pastoralists to invest the proceeds of business or paid employment in livestock kept in their home areas. But even without these trends, there are always households that have few livestock because they are newly established, have suffered some calamity, or are improvident or incompetent as pastoralists.

The effect of this imbalance on NRM depends on the extent to which societal mechanisms support needy families, whether by absorbing them into wealthier households, offering subsistence in exchange for services, or restocking deserving cases. All such mechanisms help to maintain solidarity. Their effectiveness declines, however, as group wealth declines and as social differentiation widens. Clearly poor families with just a few goats access different resources and have different priorities than wealthy families with many cattle or camels.

Mobility

Mobility is a key feature of pastoralism. Geographers were quick to differentiate nomadism and transhumance on the basis of patterns of movement (free-ranging or following set seasonal shifts). Ecologists have since emphasized the value of shorter range movements in utilizing key resource sites. However, the pattern and the range of movement describe only the superficial aspects of mobility. More fundamental to NRM is the rationale for movement.

Rationale for movement is often complex. Some movements are inescapable, when responses are by compulsion or compunction, while others involve choice, guided by the motivations and predilections of pastoralists.

Imposition occurs by *compulsion* when climatic imperatives are involved (of the type noted above under 'seasonal access'), and by *compunction* when movement is determined by customary ethics (for example, joining clan gatherings, moving with phases of the moon, or otherwise conforming to societal norms). Compunction becomes compulsion when non-compliance attracts serious social stigma or retaliation (for example, from farmers who want livestock out of their area while crops are vulnerable to grazing). Imposed movement usually benefits NRM, as when imposed transhumance ensures annual rest from grazing.

Motivation comes into play when there is a choice of where and when to move, either within an imposed framework or where no such framework exists. Movement is then determined by three or four principal motives, sometimes working in unison and at other times in opposition:

- *optimizing* livestock production and family welfare (moving so as to keep livestock output in line with family needs);
- *conserving* resources, particularly breeding stock (in order to have the capacity to recover from stock losses) but also feed resources (moving so as to rest areas for later use);
- *avoiding* hazards of disease, drought, and predation (which become compulsory movement when the hazard is great); and
- *diversifying or commercializing* production, which applies when pastoralists rely on food grain or on maximizing the sale price of livestock (when movements are organized to be near croplands or markets at appropriate times of year).

Predilection operates in ways that cannot readily be generalized. Where once the predilections of leaders would have been paramount, education and politicization now instill new allegiances. When predilection for *nomadism* wanes, more sedentary lifestyles follow. Without *conservatism* or *mutualism*, pastoralists tend to deviate from customary movements. *Opportunism* is essential for optimizing wet-season grazing areas, but is potentially antisocial in dry-season areas and anarchic when used to expand territory and stockwealth.

Implications for NRM are discussed where relevant in the main text, but four general conclusions can be drawn:

- Although mobility is the principal tool of customary NRM, movement is often due to factors unrelated to NRM.
- Effective intervention requires prior knowledge of the extent to which movement

is imposed by climatic or other imperatives, and the motives underlying other moves.

- Little is published on rationales for movement, and tapping indigenous knowledge requires more time and effort than is provided by rapid appraisal.
- The current predilections of pastoralists are not always compatible with sustainable NRM.

Resource Tenure

The concept of tenure comes into play only after rights of access have been established. Tenure refers to the procedures by which a pastoral society controls access to the resources under its control, and through which members of that society access both these resources and resources controlled by others.

This duality, allowing movement between territories, is one of the characteristics of pastoral tenure. Another is that pastoral tenure systems are defined in terms of individual resources and not of total area resources. Even if an agropastoralist or other cultivator has exclusive right to cultivate an area, that right is resource-based, applying to the soil (and of course the crops rooted therein), and not necessarily to trees growing in the cultivated area or grazing that is on offer after the crops are harvested.

How tenure relates to individual resources was explained in Chapter 1 and is not repeated here. What is relevant in the present context is how NRM is managed.

Management Systems

Customary NRM arises from a web of interacting needs, pressures, and responses. The extent to which the network as a whole comes under management is a matter of historical chance and the emergence of unifying leadership. Normally, however, pastoral societies are structured with both vertical and horizontal divisions of responsibilities.

Supreme authority is principally about maintaining ethnic identity and societal ethics, or sometimes maintaining the authority of a ruling family. But upholding pastoral ethics has many ramifications, so that a supreme authority can become involved in ruling on matters of detail, such as the ownership of an individual well.

The example most commonly cited of supreme authority exercising unifying influence over NRM is the case of the interior delta of the Niger river. By all accounts, a highly sophisticated management system was in operation (in what is now Mali) during precolonial times, whereby access to the resources of the Niger delta was closely controlled according to locality, date, and pastoral group.

Another example of unifying authority is the council meetings that still take place every few years in southern Ethiopia to maintain 'the Peace of the Boran'. Binding decisions emerge from those meetings on matters such as whether a group is entitled to preclude other users from what traditionally were kept as communal reserve grazing areas.

Area management varies from being a nebulous concept (as in nomadic societies) to being structured through two or three societal levels. In the latter case, area responsibility lies with kinship or neighborhood groups and their leadership. Unless the leadership is installed by higher authority, its character is usually diffuse, operating by consensus while acknowledging the authority of individuals in specific aspects of pastoral life. Decisions on major movements are commonly taken at this level, along with decisions about resting specific areas and controlling access by outsiders.

Individual resource management focuses principally on water. It is sometimes permissible for families to reserve grazing for their own use, but more usually this is a group decision. The group may also oversee water management, although always with an individual or family in charge of each strategic watering point. Privately-owned wells and reservoirs are the responsibility of owners or their clans, while elsewhere it is the job of a manager (sometimes hereditary) to organize maintenance and apply customary rules of access.

Women make or influence many management decisions, but their ownership of resources seldom extends beyond livestock. Issues of livestock ownership and management are often complicated. Either sex may manage livestock of the other (depending on species, location, and stage of lactation), both at family-level and when minding livestock of friends or relatives. And sometimes 'ownership' rests with the clan and not with families at all. Clan ownership extends only to the species that represent the greatest asset, usually cattle or camels.

B

External Intervention in Pastoral NRM

The freedom of pastoralists to move and operate their own systems of social organization and control has come under acute challenge this century. Colonialism and totalitarianism have both taken their toll, as has population pressure. Subsequent development efforts have not been much help either, as outlined in this annex.

Colonial Era

The effects of colonialism in Africa are of current interest for two reasons — on the one hand, these effects persist, but they also give an indication of what to expect as totalitarian and related regimes are dismantled in other regions.

Rules of tenure were introduced that were alien to local custom, and newly established boundaries restricted movement and split pastoral groups between States. Introduction of the concept of ownership led either to the State owning everything or provisions for private ownership that disadvantaged pastoralists and left them as second-class citizens. Land law and other legislation was modeled on the legal code of the relevant colonial power.

Not all was negative, however. The establishment of boundaries and their firm administration helped some of the weaker pastoral societies maintain their identity where otherwise they would have been overrun by more aggressive neighbors. In addition, discriminatory land laws were sometimes changed to improve pastoral rights.

But most legal codes have retained their original character or have been adjusted to the political character of the State, with input by lawyers trained in western or Islamic concepts and with little grounding in customary law. Meanwhile, the law has frequently been used — or perverted — to expropriate pastoral land for other uses. Agricultural settlement, ranching, and wildlife reserves have all encroached, often taking the most valuable grazing areas.

Administrative systems designed for colonial rule often sidelined or suppressed indigenous systems. This did not happen when administrators saw their principal task as reinforcing timely movement among seasonal grazing areas, but even light levels of colonial administration undermined local capabilities in management and conflict resolution. Colonial bureaucracies almost invariably proved too expensive for post-independence budgets, and too top-heavy for effective development in remote areas.

Inputs to pastoral development were modest, but the basis for present-day veterinary services was laid and a significant amount of water development took place. Research was better funded under colonial regimes than now. Systems research started later, but it is remarkable how many of the resource maps and sociological texts in use today date from the colonial period.

The colonial attitude to pastoral development was based on a mixture of paternalism and environmentalism, coupled with designs to commercialize livestock. The environmental lobby was led by soil and wildlife conservationists. The few ecologists involved drew their intellectual support from Europe and South Africa rather than from range science in North America.

Project Interventions, 1966–1993

The aid that burgeoned in the 1960s as colonialism waned brought new influences, including that of American range science. Initial inputs to pastoral areas, however, went into planning, and material assistance was slower to arrive. In the meantime, conditions were worsening — pastoral land continued to be lost to other uses, pressure on resources continued to grow, and sometimes insurgency and policies of sedentarization added to other woes. The institutional framework for development support remained inadequate, and when investment projects started in the late 1960s, they faced a daunting task.

The World Bank has been one of the main financiers. Table B-1 shows how projects have changed in their attitude to pastoral development, with the main conceptual shifts summarized at the bottom of the table.

Pilot projects were designed to extend agricultural sector lending to cover range-based livestock production. The first project was the Kenya Livestock Development Project (KLDP), which the World Bank justified at appraisal in 1967 on criteria modeled on those used for livestock projects in Latin America. Other pilot projects, similarly justified on livestock performance parameters, followed in quick succession in other parts of Africa. Some were located in the wetter zones, but like KLDP, three or four affected dry-area pastoralism.

The best way to describe these projects is that they extended government policies and plans for the national livestock industry into pastoral areas. They have since been categorized as ranching projects inspired by Hardin's 'tragedy of the commons', but in fact only one out of three arid zone projects had a ranch component. In other projects, ranching was confined to subhumid and semi-arid zones, and all were conceived before Hardin's paper (1968) became influential.

Portfolio expansion began in 1973. Second phase projects (starting with KLDP 2) and expansion into new countries meant that the number of livestock projects doubled by 1979, accompanying a general expansion in World Bank lending for agriculture. The fact that expansion took place at all in pastoral areas can be attributed to inadequate monitoring during the pilot phase.

In-depth monitoring would have shown , for example, the folly of expanding land adjudication in Kenya and increasing the number of entitled groups while the original group ranches were still laboring under inappropriate constitutions and inadequate counseling (Pratt 1990). Equally unfortunate was the expansion of ranching in Botswana and the attempt in the arid zone of northern Kenya to introduce year-round grazing in former wet-season grazing areas. The latter initiative, however, attempted through water development in demarcated grazing blocks, was the fault of planners and not of monitoring.

Not until 1975 was it accepted that project types were inadequate to deliver the expected economic benefits or contribute positively to NRM. Thereafter, from 1976 to 1979, new projects sought to establish community-based organizations that could exercise NRM responsibility without the heavy institutional trappings of the Kenyan group ranches. The livestock projects of eastern Senegal and Niger were of this type.

Retrenchment marked the next seven years, as a result of general disenchantment with range-based projects. There were few free-standing livestock projects, although there were new rural development projects with livestock components (such as in Mali and eastern Senegal), and a Botswana project that linked livestock development to land management. In addition, the Morocco Middle Atlas-Central Area Agriculture Development Project sought to establish pastoral organizations in a silvicultural/agropastoral environment to assist in range management. The few free-standing livestock projects that were started all focused on animal health services, seeking to reduce project complexity by tackling just one component. Projects of this type in Somalia and Kenya were no more successful than their predecessors, but one in the Central African Republic that relied more heavily on private initiatives was a definite success (Box 6-1).

Reentry to pastoral situations began in 1987, albeit on a modest scale. A rangeland component was included in Ethiopia's Fourth Livestock Development Project, and range-based projects followed in Mauritania and Chad. Like the Niger project of 1979, all of these focused on forming community-based organizations with NRM responsibility, although the Chad project aimed to restructure the entire livestock subsector. Most of these projects also aimed to achieve more explicit decentralization and privatization than their predecessors (Boxes B-1 and B-2). It was not until 1993–94 that the World Bank moved to adopt projects described explicitly as NRM projects.

Recent Initiatives

NRM projects are all too recent to have shown clear results. Initiatives began first in West Asia and North Africa (WANA) in:

- *Turkey*, where the Eastern Anatolia Watershed Rehabilitation Project aims to improve soil conservation and range management in 54 microcatchments covering 400,000 hectares, as part of a larger effort to protect the upper Euphrates watershed.
- *Tunisia*, where the Northwest Mountainous Areas Development Project addresses rural poverty and natural resource degradation by improving roads and social services, aiding participatory watershed rehabilitation, establishing credit associations, and facilitating planning and research.
- *Egypt*, where the Matruh Resource Management Project is providing (along with research and extension, rural finance, and project coordination) an NRM component to improve water management and the use of soil and vegetation resources in a deprived and degraded governorate in the western desert.

While these projects deal with semi-sedentary production systems, others are in line to support mobile pastoralism. There are several projects under preparation or early implementation in West Africa (Burkina Faso, Mauritania, and Niger) and one in Kenya. The Kenya Arid Lands Resource Management Project is destined to cover the northern half of the country, with components for drought management, marketing and infrastructure, and community microprojects (expanding on work initiated during a two-year Emergency Drought Recovery Project).

Viewed together, these projects show increasing commitment to providing direct support for pastoral development. Especially noteworthy is the extent to which the Kenya project is seeking to decentralize decisionmaking and involve NGOs in supporting community initiatives. The project also calls for a comprehensive review of policies and legislation covering pastoral land tenure. On the other hand, these projects are proceeding with only superficial knowledge of social-territorial organization and limited provision for systems study and monitoring. They are not process projects (Chapter 6).

Other initiatives by the World Bank include field testing of holistic resource management (HRM) under Sahelian conditions and maintaining a dialogue on other approaches. The involvement with HRM is summarized in Box B-3 and the significance of other concepts is reviewed in Annex C.

Table B-1. Approaches to Pastoral Development in World Bank Projects in Arid and Semi-arid Zones (1965-1995)

Parameter	*Early livestock and rural development projects, 1970s and 1980s*			*Current NRM projects and concepts now being operationalized, 1990s*		
Type	Commercial ranching (in countries where commercial ranching already featured)	Range/livestock development in pastoral areas	Integrated rural development (mostly in agropastoral areas) [a]	NRM projects	Opportunistic management [e]	HRM [f]
Focus	Capital investments in fencing, water, etc.; ranching technologies, private and parastatal ranches	Earlier — land rights, group ranches/pastoral units for communal areas, infrastructure (water, roads, markets...) Later — policy framework, need for mobility and flexibility in grazing rights allocation, less rigid grazing control, organization of herder-managed services	Addressing pastoral development as one component of district or regional development	Pastoral production NRM; private institutions; incentive and regulation framework; involvement of all actors	Advocates mobility, maintaining access to large areas and key resources; efficient drought contingency; better marketing arrangements	Advocates an holistic approach to resource planning, involving detailed goal setting of all range users and identification of necessary tools
Issues	Stratification of production [b], intensification through sedentarization	Increased commercialization; grazing rights allocation; herd size control	Mobilizing cross-sector inputs; delivering needed services on an area/community basis	Sustainability; NRM; livestock-crop-forest integration; traditional systems	Herder organizations, drought contingency; definition of public/private sector roles; management of cost recovery schemes	
Results	Disappointing	Limited success [c]	Disappointing	Too early for assessment	Too early for assessment	Too early for assessment
Causes of failure	Rigid organizational forms; inadequate appreciation of traditional system; assumption that pastoralism is market-driven	Inappropriate incentive framework;[d] rigid imposition of grazing and land rights; institutional weaknesses in implementing agencies	Over reliance on (a) coordinated action by (uncoordinated) ministries, or (b) creation of area development agencies, seldom sustained post-project; inadequate provision for local participation	Too early for assessment	Too early for assessment	Too early for assessment

Projects	Botswana Livestock Development projects, Kenya Livestock Development project, Yemen	Earlier — Eastern Senegal Livestock Development project, Kenya and Burkina Faso group ranches, communal area components of the Botswana projects Later — Mali Mopti Area Development project (5th Region), Niger Livestock project, Somalia Livestock Health Services project, Chad Livestock projects, Mauritania Livestock projects	Eastern Senegal Rural Development project, Somalia Central Rangelands Development project, Ethiopia Fourth Livestock Development project, Morocco Middle Atlas Agriculture Development project, China Xinjiang Area Development project	Mali and Burkina Faso NRM projects, Egypt Matruh Resource Management project, Turkey Eastern Anatolia Watershed Rehabilitation project, Tunisia Northwest Mountainous Areas Development project	Kenya Arid Lands Resource Management project, Mauritania and /Niger NRM projects, Iran Rangelands and Livestock Development project,	Sahelian sub-regional pilot program (see Box B-3)

Capital investment Blueprint planning Short-term detailed project agreements Centralized/public sector decisionmaking Livestock development Reducing stocking rates	**Shift** **1965** ——————————————→ **1995**	Institution building Flexibility(moving toward a process approach) Long-term flexible program financing Decentralized/private and pastoralist decisionmaking Sustained resource management Living with uncertainty [g]

Source: from de Haan 1994.

a. This model attracted more support from UN and bilateral agencies than from the World Bank.

b. Model in which national production is stratified into three zones — pastoral zone (zone of birth); agropastoral zone (growing-out); and agricultural zone (fattening).

c. Not successful in all aspects; some encouraging results involving herder organizations in range management.

d. Especially in WANA region through the effect of subsidies (interest rate, food supply, services, prices).

e. As outlined by Behnke, Scoones, and Kerven 1993

f. Holistic resource management (HRM), Savory 1988

g. Coining Scoones title (1994).

Box B-1. Experience with Decentralization in Rural Development Projects

The poor performance of early integrated rural development projects is now thought to be due mainly to overcentralization, with planning remote from reality, cumbersome coordination, project designs unsuited to local conditions, and insufficient beneficiary participation. These issues are being addressed in current projects by decentralized decisionmaking and financing, and improved community participation. Initial work reveals both positive and negative aspects:

Successes

- Decentralization releases latent local capacity suppressed by centralized rule.
- Communities and local authorities have proved especially capable at identifying and implementing microprojects, and improving the delivery of services.
- Other benefits include improved transparency and accountability, more effective targeting of the poor, and in many cases, improved cost-effectiveness and increased local revenue.

Remaining problems

- Resistance at the center to relinquish power and responsibility.
- Lack of a clear legal framework for decentralization.
- Inadequate funds, fiscal support, and professional staff at decentralized levels, compounded by inadequate equipment and buildings.

Significant lessons

- Decentralization requires concurrent *political, administrative,* and *fiscal* devolution.
- Decentralization fails if transfer of adequate fiscal powers is withheld, although a test period of delegated responsibility may need to precede formal devolution of power.
- Strong political commitment is needed to allow popular representation through local democratic elections and legal clarification of responsibilities.
- More research is needed to establish how decentralization can best be applied in contrasting rural situations.

Source: Based on World Bank 1995e.

Box B-2. Experience in Privatizing Services in Pastoral Areas

Privatization of services earlier delivered by government or parastatals is now well-advanced in the livestock sector, especially in animal health services. Lessons learned from past experience should contribute to the delivery of other services to herders.

Most critical when privatizing livestock services is to find the *appropriate balance* between public and private sector inputs (Umali, Feder, and de Haan 1992). To do so, it is necessary to have a clear understanding of the nature of each of the services requiring delivery. One important step is to establish the capacity of the local population, in their various social strata, to contribute to the cost of services and goods. Privatization cannot and should not be undertaken as one broad strategy.

In a policy of selective privatization, the State should:

- promote the transfer of 'pure private' goods to the private sector;
- create mechanisms to correct market failures whenever possible; and
- provide the services that are 'pure public' goods.

Because the private sector can only operate in areas with relatively high economic returns to delivery, the public sector must continue to support the delivery of many services in pastoral areas. This implies:

- *delegating* powers to local government,
- *sub-contracting* tasks to private operators and community-based organizations, and
- *subsidizing* services that are 'public goods' in nature.

Examples of sharing responsibilities between the public and private sectors according to the economic character of services are provided in Table 5-1.

Source: Internal World Bank documents.

Box B-3. Experience with Holistic Resource Management in Pastoral Development

Past attempts at checking the degradation of collectively used rangeland have been mostly unsuccessful. On the one hand, pastoral communities were barely involved or not involved at all when blueprint or rigid grazing schemes were supposed to be enforced. On the other hand, the technical principles of these interventions have often been shaky, to say the least, with overemphasis on stocking rates that have little relevance under unpredictable rainfall and mobile production systems.

The holistic resource management (HRM) model addresses these shortcomings. The model (Box 10-1) requires resource users and the concerned community to (a) establish the goals that should be achieved; (b) assess the health of the ecosystem in terms of water and mineral cycles; (c) select the tools and technologies most likely to achieve their goals; and (d) continuously monitor results and adjust management plans accordingly.

Given the growing support for this approach among ranch managers, its relevance to pastoral development was explored at the end of the 1980s by a UNDP-FAO project in six countries in North Africa and the Middle East. Despite the interest expressed by pastoral communities, little progress was made on the ground, due mostly to the reluctance of governments to transfer significant decisionmaking authority to these communities. Since 1990, Bank staff have pursued discussions in a number of Sahelian countries, with pilot perimeters of 5,000 to 10,000 hectares established in three countries (Mauritania, Mali, and Chad). Following promising preliminary results in 1994, four additional countries (Guinea, Senegal, Burkina Faso, and Niger) joined the program, and eleven pastoral pilot perimeters (PPPs) were established during the rainy season of 1995.

Preliminary results from the Chadian PPP show that resident pastoral communities have understood and are implementing the improved management plan, and have convinced temporary users (transhumants) to do likewise. Although measurement of the physical impact of improved management based upon ecological indicators will take several years to yield statistically significant data, herders are convinced that improved management under the program has already contributed to a return of grasses that had disappeared, and to a longer grazing period. The herders also believe that the management scheme has improved relationships with transhumants and agriculturists farming within the perimeter. For these reasons, they have requested a continued and expanded test.

The origins and content of the HRM model are described in more detail in Box 10-1 and Annex C.

Source: John Hall, Task Manager of the Sahelian PPPs Program.

C

New Concepts in NRM

As the World Bank became more responsive to environmental and social issues in development, these issues moved from secondary considerations in project justification to become part of the substance of its lending. These moves are still at an early stage of evolution, however, and are still in need of intellectual support in the areas outlined in this annex.

Participatory Approaches

Seeking local participation in development is not itself new. Consultation is evident even in the earliest of development projects, and few development workers now question the desirability of participatory rural appraisal (PRA) and other participation in the planning process. Where the gap remains is in allowing participation in *development*. This requires a realignment of power and decisionmaking, and compromise between what is desirable and what is manageable by local communities.

Because the World Bank now has a source book on participation (World Bank 1995d), only points specific to pastoral NRM need be considered here. If it is accepted that participation is essential to rural development and that participatory planning is pointless unless it leads to participatory development, then four points suffice:

Community participation is all important since pastoralism is a communal activity. What constitutes the relevant community, however, must be determined for each situation. Pastoral societies, as noted in Annex A, are differentiated organizationally and in stockwealth. Invariably, two or three organizational levels need to be involved and inputs varied to meet the different needs of poor and wealthier stockholders.

Societal welfare mechanisms need to be understood before any intervention is considered for the benefit of the poorer members of society. As elaborated in the main text, targeted intervention is likely to be most effective if channeled through existing mechanisms of welfare support.

Empowerment should be guided toward the groups relevant to the targeted tasks. Inappropriate links between tasks and groups may be worse than no empowerment at all. Moreover, each group must have a form of organization appropriate to the task at hand.

Mobility and the complexities of pastoral social-territorial organization cannot be unraveled by just participatory rural appraisal. Time and resources for in-depth study need to be provided both prior to intervention and in the course of project monitoring.

This is worth remembering when linking pastoral development with land-use planning and environmental management. Several countries now have schemes for *gestion des terroirs*[1] or environmental action plans at village level. When these are introduced in pastoral areas or on routes of transhumance without adequate attention to existing pastoral organization, they are liable to assign local management responsibility in a manner disruptive to the regional pastoral economy.

Holistic Resource Management (HRM)

HRM is a management package promoted by the Center for Holistic Resource Management (CHRM), Albuquerque, New Mexico. HRM is really more than a management package, but that

1. For a review in English, see Toulmin (1994).

is the mode in which HRM is most relevant to NRM — offering a means to define and operate management plans that conform both to ecological guidelines and the goals of the managers and beneficiaries.

The package already outlined in Box 10-1 also provides economic guidelines and fosters human development, but its ecological component is pivotal because HRM stems from concern for the environment and a grazing system founded on ecological principles. The grazing system is the rapid 'rotational' system developed more than twenty years ago in Zimbabwe (then Southern Rhodesia) by Allan Savory. It was experience in implementing this system that led Savory to integrate ecological resource management with precepts of holism (Smuts 1926) to form what is now HRM.

Now HRM is projected by CHRM as a way to sort out priorities and allocate time and resources in any field of human endeavor, but to most practitioners it is an approach to NRM, at the center of which are procedures and guidelines for range management. It is in this format that HRM has been most rigorously tested (see Box C-1 for an overview of the results of ecological grazing management over a twenty-year period in Namibia). It is now being tried with World Bank support in the Sahel (Box B-3). It is this model also that is elaborated in the 1988 textbook by Allan Savory (complemented by the Bingham and Savory workbook of 1990). Any resource manager who can absorb that material and has ecological appreciation and control of the area under management should be able to benefit from the application of HRM. The issue here is the relevance of HRM to pastoralism.

Merits of HRM include:

- *an holistic approach,* leading to actions that meet the broader objectives of those involved, helping to resolve conflicts

Box C-1. Ecological Grazing Management in Namibia

Under extensive grazing management on Namibia's semi-arid conditions, a new form of ecological resource use based on the HRM method has been sucessfully developed over the last 20 years. A comparison[1] of farm profitability where resources are managed traditionally (reference farms) to farms using ecological resource management clearly showed that where the HRM method is applied, both greater efficiency of resource use and higher factor productivity can be achieved.

The time series analysis at right shows that farms applying the HRM method are not only able to survive periods of drought, but also maintain productivity without any appreciable change in herd size through deliberate reduction or cattle losses (compare 1979-84 to 1974-75). Ecological grazing management enables farms to survive fodder shortages without:

- appreciably reducing herd size, since it is 'ecologically appropriate' in the long term;
- destabilizing grazing resources, since their ecological carrying capacity has been respected; and
- suffering economic losses, since herd reductions are not necessary.

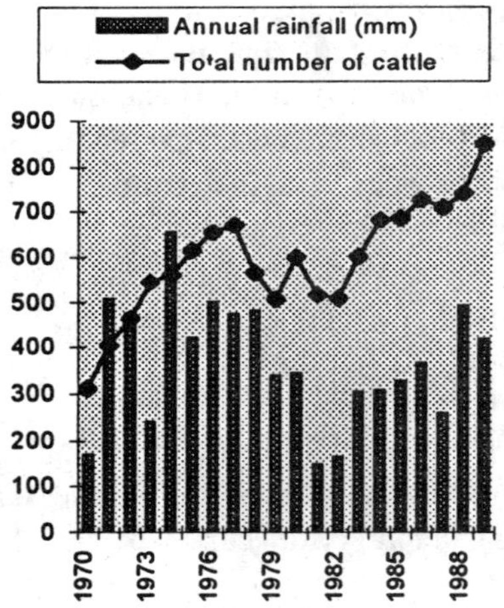

Stocking Rate Using HRM Over 20 Years

This last point is evident in the figure at right. The 1983-84 interfarm comparison shows that:

- When initial soil and climate conditions are the same, stock density is almost twice as high as on the reference farms.

and discourage 'quick fixes'based on undue dominance of one perspective over others.

- *a logical sequence* for preparing management plans and adjusting them in the light of experience (plan-monitor-control-replan is an HRM maxim); and
- *criteria*, especially ecological criteria, against which any input or output can be tested.

Limitations of HRM in pastoral situations are that:

Goal-setting becomes increasingly complicated and ambiguous as group size and interests multiply, and is barely feasible where groups overlap and survival depends on aggression.

'Biological planning' as recommended by the HRM model is an intricate process that readily attracts more external input than is conducive to pastoralists applying HRM for themselves.

Ecological guidelines currently incorporated in the HRM model are a mixture of basic ecological truths and guidelines specific to the ecologies with which Allan Savory is familiar. Some situations are not covered, such as semi-desert and camel-based systems, and environments characterized by cyclic succession between bush and grass or between annual and perennial grasses.

Conclusions about the use of HRM in pastoral development include three points:

First, although small and homogeneous pastoral groups may have no difficulty in setting goals, success cannot be expected if customary mobility is restricted in order to create a more manageable situation. If external advisers make management decisions, the main merit of HRM, as perceived by its founders, is lost.

Second, the ecological limitations are largely self-inflicted by the construction of the HRM

- The management system reduces veterinary costs (by 80 percent) and eliminates the need for mineral supplements.
- Productivity of the land in terms of gross income per hectare is well above that of the reference farms (3.7 times higher) because of the stocking density.
- Capital investment (machinery, water supply equipment, fencing, and pasture improvements) is more than 50 percent higher than on the reference farms, reflecting changes to allow for paddock rotations.
- The capital-intensive paddock technique reduces labor input (52 percent lower labor costs per animal compared to reference farms).
- The return on total factor input (grazing resources, equipment, machinery, and labor) is far higher on the ecologically managed farms (1.18 Rand/ha) than on the reference farms, where it was negative after 6 years of drought (-0.94 Rand/ha).

1. The study is based on a database of several years, a rarity in developing countries.

Source: Otzen 1989.

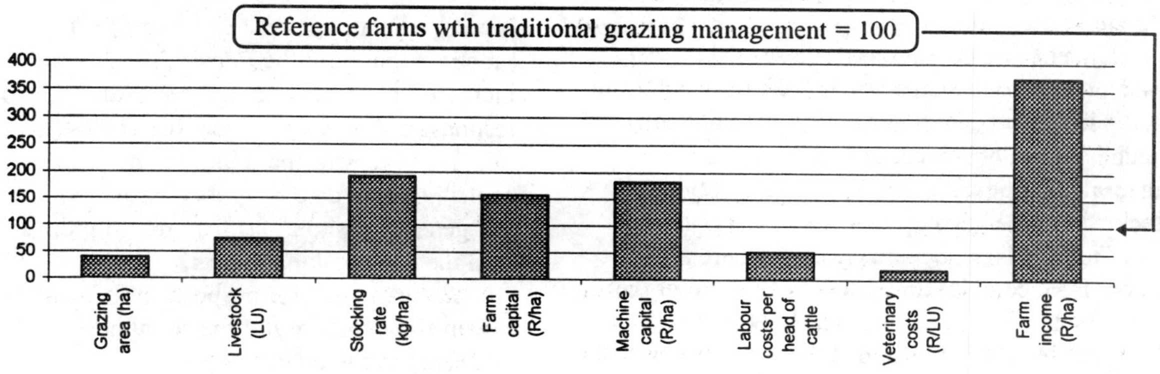

Comparison of Reference Farms to Those with Ecological Grazing Management model, which compresses the whole variability of world environments into one scale of brittleness. A parallel ecoregional categorization would be helpful, so that new thinking (see below) and new research could be incorporated, as appropriate, in guidelines for each specific ecoregion.

Third, some people claim to use HRM when they are merely using a rapid rotational grazing system. Moreover, the grazing system is often applied in a prescriptive form, with fixed grazing periods instead of plant growth monitoring to adjust rest periods, as the HRM model requires. *The point of HRM is not to follow prescriptions, but for resource users to use the HRM model to make their own decisions.*

Rethinking Range Ecology

As American-trained ecologists began to influence pastoral development, it became evident that much of formal range science was inappropriate in that context. Written commentary grew during the 1980s and led to the workshops of 1990 and 1993 that produced *Range Ecology at Disequilibrium* (Behnke, Scoones, and Kervan 1993) and *Living with Uncertainty* (Scoones 1994). Both books focus on pastoralism in Africa.

The thesis is that pastoralists do not habitually overgraze their land. On the contrary, the intensive use to which they subject rangeland represents efficient range management, given (a) the management objectives of pastoralists, (b) the extreme variability of rainfall and grass production, and (c) the capacity of African rangelands to recover, with rainfall, from a depleted state.

Under those conditions, the use of the concept of carrying capacity as a decisionmaking tool is irrelevant, if not dangerous. It might suit ranching, but not African pastoralism. Under pastoral conditions, mobility and flexibility in stocking are the key requirements.

This thesis is not entirely new. There are papers from colonial times that make some of the same points. Gilles and de Haan (1994) also show that World Bank-funded projects in West Africa already contain provision for mobility and flexibility, even though the projects are not expressed in terms of 'opportunistic range management'. The new literature, however:

- provides a theoretical base by differentiating rangeland systems that exist in a state of relative equilibrium (the ranch model) from those characterized by disequilibrium (where much of pastoralism is practiced), and
- elaborates implications for pastoral development with particular reference to pastoral systems that retain a capacity for free movement.

In addition to the two books cited, a useful summary appeared as the first issue of the Overseas Development Institute's *Natural Resource Perspectives* (Behnke and Kerven 1994).

The main value of the contribution to date has been to highlight the merits of pastoralism and the need to approach pastoral development differently from ranch development. Pastoral strategy tends still to be cited as 'opportunistic', but now with the useful distinctions between:

- *tracking*, meaning 'prompt realignment of livestock forage demands with fluctuating levels of primary production', and
- *buffering*, meaning 'shielding of pastoral incomes from the worst effects of the climatological and biological roller-coaster' (Behnke and Kerven 1994).

'Absorbing' could also be recognized as a third category of response to cover situations where pastoralists absorb the vagaries of an arid environment by keeping physiologically adapted breeds, adjusting family diet, or following practices that limit the rate of population increase. It seems better to keep these types of response separate from the other two.

However, the main limitations of the contribution to date lie in its tendency to polarize and generalize situations that are far more subtle than the new thinking allows:

Polarization, for purposes of emphasis, has sidelined the majority of the situations which planners need to address:

- Ranching vs pastoralism acknowledges ranching only in its confined and commercialized mode and assumes that pastoralism is still always freely mobile.
- Equilibrium vs non-equilibrium makes two disparate states out of what in reality is a continuum (and devalues succession as a factor in non-equilibrium situations).
- Degradation is defined so that only irreversible change rates as important (while human degradation is disregarded altogether).

Generalization has compounded polarization, specifically:

- The resilience of African rangelands does not apply to all grassland types, and especially not to browse. When browse plants are killed by overuse, it takes not one good season to restore productivity, but more usually ten to twenty years in the semi-desert, where browse is the main feed supply.
- Caricaturing all past development as 'ranching' obscures the other problems that must be recognized if new thinking is to produce results any better than the past.
- Idealizing pastoralism obscures the suffering that opportunistic management brings to pastoral families during drought — the society recovers, but not the families that have been decimated or forced out of pastoralism in the process.

It is not helpful that the new thinking perceives rainfall as external to pastoral production (when pastoralism is itself a response to rainfall). The model also largely ignores time as an ecological factor — although periodicity in rainfall and ten- to twenty-year events are enormously important in maintaining diverse vegetation. In terms of NRM-oriented development, 'opportunistic management in non-equilibrium situations' is appropriate to mobile pastoral systems that occupy moderately productive grasslands on dissected landscapes within an arid to semi-arid environment. It is not well-suited to semi-desert or to areas with highly erodible soils, and is limited under other conditions as summarized in Table 3-2.

In the future, more use should be made of chaos theory in seeking to explain the behavior of arid zone systems. This is not the place to elaborate on chaos, but it is now a recognized field of study, a science of process rather than state (Gleick 1993). Chaos is more likely than loss of equilibrium to characterize arid zone pastoral systems. The notion of aridity increasing chaotic behavior is certainly more plausible than the notion of equilibrium and non-equilibrium existing as disparate states, with the latter somehow washed of the physical, physiological, and temporal processes that drive plant succession.

D

Features of Pastoral Systems

This annex elaborates on Chapter 2 by describing each of the pastoral systems cited in terms of their environment, territoriality, and livestock, and by outlining relevant development interventions and the nature of allied systems.

Mobile Pastoral Systems

Mobile pastoral systems are forms of pastoralism in which family herds or flocks, and often the families themselves, transit tens or hundreds of kilometers each year, shifting every few days or weeks. The movement may be nomadic if unpredictable and without fixed abode, or it may involve transhumance on a regular seasonal schedule between fixed localities. In between is peripatetic movement — not fixed in its timing or direction, yet not random either.

Desert Camel Herding

Environment of the 'desert' includes semi-desert, and it is there that most activity is concentrated. Mean annual rainfall is less than 200 millimeters, and plant productivity is due as much to water flowing into the zone as to incident rainfall. The vegetation of water courses, particularly browse, is the basis of animal subsistence. Flushes of annual grass that follow rare rainstorms are often wasted, and the annual flora provides no guide to grazing or population pressure. Dead perennials may indicate overuse, although dry cycles and changes in water course configuration also cause trees and shrubs to die.

Territoriality focuses on the resources of water courses. Each drainage system is commonly associated with a kinship group that ranges up and down the one system. Individual wells can assume enormous strategic value and may be defended most vigorously if unauthorized use becomes known to the owners. If there are sufficient oases or wells, pastoralists may stay year-round in the semi-desert, although retreat elsewhere is unavoidable if browse supplies fail.

Livestock are principally camels because only this species can simultaneously:

- utilize the feed on offer (from grass to tall trees),
- traverse dunes and the long distances between water points,
- be a productive milk animal under these conditions, and
- transport all household chattels when families move.

As a result, the camel has a high sale value within and among desert communities, although the size and value of an animal means that it is an extravagant source of meat. Partly for this reason, goats (or sheep in central Asia) are kept as a supplementary species. Small stock are also readily marketable, whereas in some countries urban demand for camel meat is low (Table 2-2).

Development interventions are likely to focus on maintaining existing territoriality and easing water lifting and perhaps adding extra wells. The training and equipping of mobile health auxiliaries to serve both people and animals would also receive early attention, perhaps followed by provision of service centers at strategic points. Specific inputs to NRM would focus on managing browse stocks, agreeing on a drought strategy, and diversifying oasis agriculture where feasible.

Actual inputs would, of course, reflect local perspectives. Local variation to be considered includes the different drought expectancies in monomodal and bimodal rainfall areas, and the trade-related issues noted below. 'Rethought' range ecology (Annex C) has little relevance to these areas because of its orientation to grass.

Allied systems arise when subsistence systems become monetarized. Desert camel herding is usually milk-subsistent in the tropics, but more trade-oriented where Bactrian camels are kept. Both the Bactrian camel and sheep that accompany them have coats appropriate for a cold desert climate, so surplus hair and wool can be sold, as well as sheep. Apart from that dichotomy:

Investment-based herding is becoming more common as non-residents invest proceeds from employment or business in livestock. The practice is more common with cattle in dryland areas, but extends to desert camels also. These animals are not always obvious because they often run with residents' herds, but the system is, nonetheless, distinct. The primary role of the animals is to combine owner satisfaction and capital growth with longer-term social security. Although the animals add to total stockwealth and provide subsistence for their herders, they also reduce the feed available to residents' herds.

Motorized desert herding is now quite common in oil-rich countries. This system also involves investment, but with the money invested in the procurement and operation of a vehicle. The vehicle may be used to transport drums of water, and even camels, to waterless areas where there is good grazing. Alternatively, feed may be taken to areas where there is water but no grass or browse. The system is now attracting milk-subsistent families with just enough animals to convert a few into a truck to manage the rest.

Dryland Shepherding

Dryland shepherding describes the most widely distributed pastoral system of the non-desert areas of the Middle East and central Asia. Sheep and goats are the principal species, with no more than a few larger animals kept for transport.

Environment is relatively uniform — dwarf shrub grassland formed under an arid climate with monomodal rainfall and a distinct cold season. The main variation is topographic, with colder uplands and cultivated (often irrigated) valleys.

Territoriality is conditioned by the cold season, involving transhumance between higher elevation summer grazing areas and lower elevation winter areas. Where there is no marked transhumance, movement is more peripatetic than truly nomadic. Movements are also conditioned by the need to access markets in order to sell livestock and purchase grain and other commodities. Dryland shepherding is thus often trade-dependent.

Livestock are sheep and goats. Sheep are used for subsistence and sale. Although meat is the staple diet, grain for the accompanying bread usually must be bought, as is the case with most meat-based subsistence systems. Milk products supplement the diet, and wool is used for home weaving. Sometimes cheese and rugs are sold. The wool is mostly sparse and coarse, although some pastoral groups specialize in breeds with higher-quality fleeces or pelts.

Goats are a useful complement because of the different attributes of their meat, milk, and coat. They are also useful because of their more inquisitive and sharper nature — when in a mixed flock, they lead sheep to succulent pasture and give warning of impending danger. Transport animals are usually whichever equine species is most readily available.

Development interventions have better prospects than in many systems, but need a framework that gives priority attention to:

- social organization and present social services;
- trading options as conditioned by dietary practices; and
- NRM options in the individual parts of the grazing orbit.

Range improvement is relatively easy in terms of overseeding and establishing fodder shrubs, but management of improved areas is always difficult in mobile pastoral systems. But since shepherding proceeds at a slower pace than herding and needs less water at each supply point, it is easier to control access and affect NRM with the help of small, strategically-placed, water storages. Peripatetic educational and health programs should also be easier to organize in this system than in many others.

Allied systems involve variants of breed type and extremes of aridity or cold. Obviously systems that use breeds noted for the quality of their hair, wool, or pelt offer opportunities to improve the production and marketing of those items. A more distinct system, important in central Asia, is where horses are the principal species and sheep are secondary:

Horse-based systems are mostly the steppic equivalent of the milk-subsistent cattle herding discussed next. Where horses are the centerpiece of the pastoral economy, the system is usually a relic of a former militarily mobile system. The horse is also particularly well-adapted to temperate uplands, and still retains the advantage of allowing pastoralists to keep in touch socially and with trading centers across long distances. The trading element of the system is as likely to derive from the sale of sheep as from the sale of horses.

Milk-Subsistent Cattle Herding

Many of the pastoral systems of East Africa are of this category. Zebu cattle are the basis of the economy, although with highly significant contributions from accompanying small ruminants.

Environment influences NRM and development to a greater degree in these pastoral systems than in most dryland systems. This is partly because of environmental variability, but also because the systems are milk-based. Because milk-subsistent systems must try to keep cows in milk year-round, each range type must be used when it best contributes to maintaining milk supply. These systems are therefore especially vulnerable to loss of key resource sites. Most of these areas can support perennial grasses, but vary greatly in value and accompanying woody vegetation.

Territoriality reflects the points just made. Some cattle-herding systems are confined to the arid zone, but most occupy or overflow into the semi-arid zone and several have access to wetter areas for dry-season or drought use. Movement is normally peripatetic and not fixed to one pattern of transhumance. Determining the basis of movement and reciprocal access is a precondition of effective intervention. Each pastoral society and group must be assessed individually.

Livestock comprise cattle, sheep, goats, and sometimes a donkey or two as a load-carrier. The small ruminants are kept principally for meat, and as 'small change' for purchases and minor social exchanges. Sometimes they are milked, but most milk comes from cattle. Herd composition is geared to milk production (although because this is the same composition needed to rebuild herds from heavy stock losses, it is not significantly different from that of other cattle-based pastoral systems).

Development interventions must be preceded by analysis of social-territorial organization, with particular reference to drought strategies and specific strategies for helping to maintain milk production. Animal health, as always, is an important area for support, but there may be little value in seeking to improve livestock marketing, except to facilitate the offtake of small ruminants where these are abundant. Opportunities for breed and range improvement depend entirely on local conditions. Key sites warrant priority attention, as does wildlife potential sometimes.

Allied systems lie in three directions — toward trade-dependent or investment-based cattle herding, and toward using camels instead of cattle for milk-subsistence in the drylands. The investment-based system has already been described in the context of camel herding. The trade-dependent situation and the overlap between camels and cattle in milk-subsistence are considered below:

Camel herding is not confined to deserts, but overlaps with cattle herding across the whole of the Sahel and through East Africa into the Arabian peninsula. In addition to camel herders extending into the drylands, milk-subsistent cattle herders sometimes keep a few camels. While the two species are complementary in their feeding habits and several other respects, conflict often arises between the two societies when they overlap. Also, in Oman, cattle herders complain that concentrations of camels spoil the grazing for cattle. The persistence of camel odors and effects of dung, however, have not been investigated.

Trade-dependent cattle herding arises where pastoralists or herding systems become dependent on purchased goods. The most typical situation is where pastoral groups adopt grain as a dietary staple while not themselves cultivating. This is the practice over much of West Africa, where trade for grain is aided by transhumance into agricultural areas each dry season, a pattern that also has environmental logic because the latter areas are subhumid and unusable during the rains without tsetse control or trypanocidal drugs. Prospects in trade and crop-livestock complementarity open a whole set of development options not found elsewhere. There is also scope to use fire and seeding in the subhumid zone beyond that found in East African systems or in the Sahel.

Agropastoral Transhumance

Environment is important in agropastoral systems because the comparative advantage of these systems hinges on successful cropping, which requires a favorable site. Sometimes the cropping area is in a zone of higher rainfall, but often lies in a flood plain or run-on area within the arid zone. The direction of transhumance also varies, towards a wetter or a drier zone, and may change over time as competition increases in outlying areas.

Territoriality is more easily described. Usually the area where the cropping takes place constitutes the homeland, and grazing areas on the route of transhumance are shared with other users. The timing of movement is geared to the needs of the cropping component — work oxen need to be on hand for plowing, and stock should be elsewhere when growing crops are susceptible to damage. Some animals stay near at hand, however, to ensure a milk supply for those who remain at home.

That synopsis applies where the pastoral group maintains a year-round presence in the cropping area. There is another type of agropastoral transhumance, where crops are planted at a suitable site on the transhumance route and are left untended until the group returns at harvest time. This system is less common than it was because of the risk of damage to untended crops.

Livestock include a mix of cattle and small stock, although with more male animals than usual in the herd to ensure a sufficient supply of work oxen. There is also advantage in having a milking type of sheep or goat so that fewer milk cows need to be held back when the cattle leave on transhumance.

Development interventions should give particular attention to the cropland, including the nutrition of oxen. Cropland provides opportunities for technical innovation not usually associated with pastoral development. Greater net benefits are likely to accrue from focusing inputs in that direction than from spreading the same resources over the animals and areas involved in transhumance. That said, keeping transhumance routes and options open is critical to the viability of the whole system.

Semi-Sedentary Systems

Before continuing with the description of specific systems, three general features of semi-sedentary systems warrant stress:

Loss of mobility. Semi-sedentary systems continue to be formed as mobile systems lose mobility. This may arise because:

- pastoral territory is lost to other uses;
- the State has a deliberate policy of sedentarization;
- people prefer — and new wells allow — fewer moves; or
- loss of stock wealth enforces reduced mobility.

Depending on the causes, different systems may appear. In the last case, the resulting semi-sedentary system arises *within* the mobile system, practiced by families with fewer and less mobile livestock than those who move freely.

Systems that are naturally immobile. There are also systems that are semi-sedentary because there is no need to be mobile. Usually this situation arises in zones that are subhumid or wetter, but it also happens in arid zones where there is a reliable source of dry-season feed from irrigation, a processing industry, or marine resources. Dried sardines have supported pastoral systems bordering the Arabian sea for many generations.

Preconditions for intervention. Distinguishing the causes of reduced mobility is essential for effective intervention in semi-sedentary systems. In general, development should aim to assist mobility, but where there is no need to be mobile or a poverty trap is involved, semi-sedentary systems warrant full support.

Goat Herding

Semi-sedentary goat herding probably engages more pastoralists and agropastoralists than any other system, at least in Africa. The system could be termed 'resource-poor goat herding', except that the implication would be that the goats are resource poor! The system is practiced both by resource-poor pastoralists within cattle-based systems and by agropastoralists who have no cattle.

Environments for goats should be arid or semi-arid with abundant browse. The only way that goat-based systems can fulfill their primary function of supporting needy families is to have access to sufficient browse to keep the goats productive when sheep and cattle are suffering from dry-season stress.

Territoriality is prescribed by the need to keep the means of family subsistence close to the homestead, yet there is also a natural constraint to goat mobility. Although agile and capable of roaming far, goats do not tolerate environmental change. If faced with unfamiliar browse or moved to an unaccustomed damp environment, mortality can be very high, with disastrous results for a family dependent on just a few animals. A further factor is that a goat-herding family is likely to be a small family, with insufficient labor to engage in long-distance treks.

Livestock may include sheep as well as goats, but not other species except possibly a house cow and one or two donkeys. Donkeys are strategically very important in agropastoral situations, since even one can help cultivate a larger area than would be possible using hand labor alone. The area difference may be critical to family survival.

Development interventions are likely to give priority to improving goat husbandry through housing and health-care. Where high-quality browse is scarce, there may be scope for planting additional species to widen the spectrum of browse plants. Planting is always more manageable in a semi-sedentary system than a mobile one.

Institutionalized Herding

Pastoral systems that operate within an institutional framework prescribed by the State cannot be generalized in terms of environment or livestock enterprise. They are located wherever the State sees fit to exercise its influence, and they start with whatever livestock existed before the new institutional framework was introduced. What they do share, however, is circumscribed mobility and decisionmaking.

There are two principal types of institutionalized herding — one the result of the central planning associated with totalitarian governments, and the other the result of extending land adjudication into pastoral areas. Both are outlined here.

Centrally-planned herding is now in decline, following the liberalization of regimes released from the control of the USSR. The aftermath remains, however, in both Asia and African countries that chose a socialist path for a period following independence. And several pastoral systems still have politicized leadership structures and work to targets set by the State. In such cases, development inputs need to be conceived within a strategy appropriate to the political environment. Initial emphasis is likely to be on reforming legal instruments and local organizations, although conditions created under totalitarianism may favor technical intervention and commercialization, provided that the area is not too arid or movement too constrained.

Group ranching is the system that most people associate with land adjudication in pastoral areas. The result of land adjudication, however, depends on how the law is written, how people present themselves for adjudication, and what follow-up is provided. In Kenya, where the idea started, the law allows any size group, from a tribe to a family, to be granted freehold title to the area to which they can demonstrate customary right. There is also provision (unused to date) to register rights not amounting to ownership. The result of land adjudication in Maasailand — although not yet in other parts of Kenya — has been subdivision and trade in land, leading to two conclusions:

Land adjudication is a radical intervention that is best reserved for controlling land grabbing in medium- to high-potential range areas. It would be wrong to withhold land title from pastoralists if other citizens have that right, but in arid areas procedures need to lead to large allocations and associated organizations based on customary authority and decisionmaking.

Development interventions in areas where land title has already been granted need to focus on helping pastoralists improve NRM on their existing holdings and avoid ill-advised action that jeopardizes future development.

Commercial Ranching

Ranching describes the practice of running or exploiting animals on a demarcated area of rangeland that is privately owned or allocated for that purpose. The land may be held or allocated under freehold or leasehold title or in the form of a legal right of occupancy. The animals may be wild or domesticated (hence game ranching, cattle ranching, etc.) and they may belong to the title holder, a tenant, the State, or someone who has rented grazing or otherwise secured right of access.

Unless the title or certificate of occupancy rules otherwise, the resource users are under no obligation to stay confined to the area allocated, exclude or allow other users, or maximize profits. Consequently there can be many types of ranching. Commercial ranching is the type that *does* operate for profit. It is not practiced much in pastoral areas, although some group ranches may move in that direction, as may cooperatives and other forms of State-sponsored enterprise in pastoral areas.

Environment for ranching extends over a wide range of climates and vegetation types — into the subhumid zone where there is available land and to the semi-desert for karakul sheep ranching in southern Africa. Most ranches, however, occupy the middle ground, with relatively productive perennial grassland or savannah in the semi-arid zone.

Territoriality has three dimensions — the size of the ranch, the extent to which the ranching enterprise is confined within its boundaries, and the NRM practices that are adopted.

Seldom are ranches of a size that is sufficient to meet management objectives in all years. For this reason, it is normal ranching practice to engage in agistment, sending livestock elsewhere when grazing is inadequate, and renting out ranch land to others when the available grazing exceeds needs. Breeding ranches, where it is preferred to keep valuable breeding lines free from disease hazards, may seek to avoid agistment, but that is the exception rather than the rule. The practice of agistment is detailed here because commentators on group ranches in Kenya mistakenly thought that movement between group areas in times of drought was evidence that ranching was not working.

There are other points of similarity between ranching and pastoral practice. It is quite normal on ranches to manage key resource areas just as judiciously as in pastoral areas. Of course, a ranch is unlikely to have the same variety of key resource sites as a pastoral area one hundred times larger, but it is still possible to embrace substantial variety in an area of 20,000 to 200,000 hectares if the topography is right. And many pastoralists are almost as restricted in their mobility.

Livestock on commercial ranches are chosen considering:

- climate, vegetation, and disease hazards;
- ranch infrastructure;
- operator's preferences for breed and enterprise; and
- market forces and available support services.

It is unnecessary here to consider all the permutations. Suffice to say that the outcome — the commercial success of a ranch — is a product of choices, management skills, and the luck that graces the chosen enterprises.

Development interventions will be different when starting ranch development from scratch than when aiding existing ranches. In the former case, a heavy input would be required to determine social and ecological feasibility, the units from which ranches would be formed, and legal instruments through which land would be demarcated and held. Commercial ranching is likely to be appropriate only where the existing pastoral system is semi-sedentary, trade-based cattle herding, with exclusive use of an adequate area of productive semi-arid or subhumid rangeland.

E

Participatory Rural Appraisal (PRA) Methodologies

PRA is a collaborative community-based decisionmaking procedure that supports participatory development. It evolved from rapid rural appraisal (RRA). While the latter enables development practitioners to collect and analyze data relevant to rural development, the former is based on data collection and analysis by local people, with outsiders acting to facilitate rather than control. The use of PRA enables development practitioners, government officials, and local people to work together on context-sensitive programs.

Methods and Tools

PRA is an exercise in communication and transfer of knowledge. The learning-by-doing and teamwork spirit of PRA require transparent procedures. For that reason, a series of open meetings generally frame the sequence of PRA activities. The following non-exhaustive list of techniques illustrates the range of methods available.

Enquiries are by observation and interview:

Direct observation is an essential first step for outsiders to learn something of local livestock management and production through first-hand observation and recognition of key indicators.

Transects are systematic walks taken with key informants through an area to give an overview of the production system and natural resources.

Indigenous knowledge is the key source of information about local production systems.

Semi-structured interviews complement other participatory methods by using a flexible interview guide to explore issues and generate discussions that may lead to more visual techniques.

Case histories and animal biographies provide data on livestock output, mortality, and fertility by systematically recording information from herders.

Ethnoveterinary question lists provide details of local perspectives on animal diseases.

Visual aids are used in both capturing and analyzing information:

Social and wealth mapping locates and records the distribution and circumstances of households and the superficial social features of the area.

Opportunities and services mapping investigates the availability of animal health services, marketing opportunities, reserve grazing areas, etc.

Resource mapping indicates which natural resources in the area are used by which livestock at what season of the year.

Mobility mapping provides a representation of where, why, and how often people travel, either with or without their livestock.

Anatomical body mapping is used to help understand local knowledge and perceptions of animal anatomy and physiology and the effects of diseases or treatments.

Systems analysis diagrams prompt discussions on the details of the livestock production system (inputs/outputs, services, opportunities/constraints/solutions, etc.).

Process/flow diagrams summarize sequences of events (such as production operations or daily activities).

Venn diagrams (*institutional maps*) explore the relative importance of services and institutions to a community.

Network diagrams investigate the different networks surrounding livestock production.

Decision trees help to identify the range of strategies available to livestock producers.

Livelihood analysis encourages people to consider income and expenditure in relation to goals and past and present coping strategies.

Problem and solution diagrams draw perceived and actual solutions for problems faced by producers.

Change over time can also be presented visually:

Seasonal analysis calendars establish how seasons are defined and indicate the distribution of activities and trends over the year and across longer time periods.

Activity profiles explore typical activities and routines, including livestock and household duties.

Time lines and trends plot significant past events and show changes that have occurred in a community.

Historical maps and transects are used to explore resource and social change over time (past, present, and perceived future).

Historical matrices help further to understand community livelihood and coping strategies, past and present.

Ranking is necessary for several attributes:

Wealth ranking groups the community into different wealth strata to show how attitudes, decisionmaking, and production priorities are affected by wealth, and to provide a baseline against which the impact of future interventions can be measured.

Preference ranking and scoring ranks people's priorities, action, and interests, either using a game board or by other scoring techniques.

Matrix ranking and scoring enables a range of different items to be assessed against selected criteria.

Proportional piling establishes informants' perceptions of relative proportions by using local natural materials built into piles.

Organization

A typical PRA involves a team of people working for several weeks or months (according to the scope and study area) on workshop discussions, fieldwork, and analyses. Techniques can be combined in a number of different ways depending on the topic under investigation. Mapping and modeling are good techniques to start, while wealth ranking is best done later. The current situation can be shown using maps and models, but subsequent seasonal and historical diagramming can reveal changes and trends. Preference ranking is a good ice-breaker at the beginning of a group interview and helps focus the discussion. Later individual or small-group interviews can follow up on different preferences and explore reasons for these differences.

Specific Requirements for Pastoral Development

Literature on participatory enquiry has mushroomed since the late 1980s, but most work focuses on crop farmers. Experience with participatory pastoral planning, however, has recently been reviewed by Waters-Bayer and Bayer (1994).

Specific needs in pastoral planning. Participatory planning in pastoral settings differs from planning in other agricultural settings because of:

- mobility of animal assets;
- variability and unpredictability of forage resources;
- territoriality (wet- and dry-season grazing and emergency reserve areas);
- dependence on common property and evolving multiple resource uses and users (requiring negotiations to access, manage, and improve the resources); and
- flexible decisionmaking (not rigorously institutionalized collaboration among independent basic operational units).

These characteristics highlight the primacy of institutional over technical concerns in pastoral development planning. For these reasons, *conflict management, pastoral organization*, and *land-use planning* are of greatest interest. PRA techniques can help to create forums to discuss and negotiate these issues.

Field experience, while growing, is still weighted heavily towards the initial stages of planning rather than to later stages and post-planning activities.

Participation in initial planning (situation analysis) is evident across a wide range of situations and countries:

- PRA methods have been used to assess forage resources (Nigeria, Ethiopia, Zimbabwe) or plan animal health care projects (Afghanistan, Tibet, Somaliland, Kenya) using methods that give a quick overview of factors affecting productivity.
- In Mongolia, government administrators and researchers are being trained in PRA to help pastoralists evaluate their own situation and familiarize officials with pastoralism. Elsewhere (for example, Tanzania) seminars have been held to sensitize policymakers to pastoral land use.

Participation in later stages (in planning, implementation, and monitoring) is evident mostly in old projects started before PRA was 'in vogue':

- In Kenya, consultations in the field with influential Maasai and herding families helped shape the concepts and legislation that formed the basis of group ranch development (although the later stages of implementation proceeded without such input).
- Also in Kenya, a series of participatory workshops was organized by the Forestry Department to agree on action to protect the natural vegetation. This approach helped to guide forest policy, including drawing up new legislation to integrate traditional rules and modern laws.
- In Mali, herders were encouraged to draw up a drought management plan of action, implement their plan, evaluate the results, and plan further activities. Other herders and farmers are negotiating joint management of natural resources, with frequent monitoring and renegotiation in response to changing conditions.
- Also in Mali, an NGO working with the Tuareg designed a self-evaluation approach based on GRAAP methods.[1]

Issues and lessons for donors. Application of PRA requires considerable skill and sensitivity to avoid pitfalls:

Communication. Facilitators need both to listen and ask relevant questions, avoiding an extractive way of applying PRA techniques. PRA visualization methods help overcome language barriers, but people who speak the local language and understand local 'codes' must be well-represented in the PRA team.

Inequity. Superficial PRA can easily allow small powerful groups to dominate the process. PRA should involve an analysis of differences among social groups, with planning activities organized accordingly, as with the Borana in Kenya (Swift and Abdi 1992).

Concepts. Confined views of location and time can readily ignore seasonal resource users. PRA is best applied by people working continuously, or at least repeatedly, in the project area. Certain PRA techniques impose foreign concepts not readily understood in all pastoral societies (for example, pastoralists in Afghanistan with a strong verbal culture had difficulty expressing themselves in diagrams, and Tuareg in Mali explicitly requested a shift from visualization to the written and spoken word for assessment and planning activities). In mapping, concepts of space based on villages with clearly defined boundaries may prove alien to pastoralists and require a conceptual shift to focus on centers of activity such as key resources and movements.

Quality. The rapid spread of PRA methods in recent years has revealed a major problem of quality control. To permit judgement on the

1. GRAAP stands for *Groupe de Recherche et d'Appui pour l'Autopromotion Paysanne*. ACCORD has produced documentation on this process, giving an overview of the strengths and weaknesses of monitoring and evaluation applied over six years (Capezzuoli 1994).

quality of the data (particularly judgement of the degree of participation by local people), documentation must describe how these methods are applied and by whom.

Ethics. In promoting participatory planning, donors should be clear about the ethical issues involved. In particular, they should accept that participatory planning must be followed by appropriate follow-up. Since pastoral practices may differ from state policy, care must be taken that the information derived from PRA is not used primarily to strengthen central power.

Review of experience with PRA in pastoral development has led to the set of recommended techniques indicated in Table E-1. For more details on PRA techniques, also see the case studies in IIED (1994) and Waters-Bayer, Bayer, and von Lossau (1995).

Table E-1. Recommended Methods and Tools for Participatory Planning in Pastoral Situations

Type of information or purpose	*Methods suggested*
Planning phase and establishing rapport	
History of area (past trends, accomplishments)	Timelines, oral history
General information on area and production conditions	Transect walks, participating in daily tasks
General information on people and relationships	Listen and learn[a]
Situation analysis (first-round enquiries with key informants and at group meetings)	
Relative importance of livestock in livelihood system	Livelihood analysis, proportional piling
Resources available to livestock	Seasonal resource mapping[b]
Resource use	Bioresource flow diagram, mapping,[b] proportional piling, matrix
Grazing pattern/forage resource use	Calendars, resource use mapping[b]
Fodder preference	Ranking[b]
Animal husbandry practices	Seasonal calendars, mobility maps
Local knowledge of livestock diseases	Ethnoveterinary guide,[b] causal diagram
History of livestock diseases	Timelines
Preferred traits of livestock	Matrix scoring
Relative mortality in different species/age groups	Proportional piling
Livestock productivity parameters	Progeny histories, herder recall
Livestock linkages with other sectors	Flow diagram
Seasonal trends, for example, disease and parasite load; mortality of livestock; livestock sales and prices; prices of inputs products, items needed; birth events in livestock; milk yield	Calendars, proportional piling
Proportional income from livestock products	Proportional piling, diagramming
Labor requirements	Seasonal calendars, daily timelines, learning local tasks
Stock loaning and sharing relationships	Social mapping
Social organization	Venn diagram, social mapping[c]
Institutional links	Venn diagram
Wealth differences	Wealth ranking
Marketing structure	Flow diagram
Conflict analysis	Venn diagram, flow diagram, critical incident

Table E-1. (continued)

Type of information or purpose	*Methods suggested*
Situation analysis (first-round enquiries with key informants and at group meetings) **[continued]**	
Innovation history	**Pathway diagram charts**
Services available	**Venn diagram, services and opportunity map**
Problem analysis	**Problem tree, causal diagram**
Detailed planning (deciding development inputs through group consensus)[d]	
Prioritizing problems	**Brainstorming /ranking**
Prioritizing solutions	**Brainstorming /ranking, problem and solution game**
Allocating tasks, time planning	**Process diagram, matrix**
Deepening situation analysis	**Same tools as situation analysis but more topical**
Monitoring and evaluation (meetings/workshops to review progress)[e]	
	Ranking and scoring techniques, series of calendars and/or maps, impact diagram

a. The source recommends taking photographs and giving and discussing prints. This can be useful, but in some societies would cause offense.

b. These methods require as a first step the collection of vernacular names for localities, plant species and parts, and other natural resources, including categories and conditions of livestock production. Uses external to livestock production (in human health and household economy) should also be covered.

c. Also in-depth discussion with elders and others, encouraging elaboration of their procedures in comparison with those of other societies known to them.

d. Semi-structured interviews can be used in all stages of planning, but are particularly useful here, when discussion needs to be guided toward decisions. Decisions should be made mostly by group consensus, although for purposes of deepening situation analysis, it is necessary to identify individuals who are recognized authorities in the subject areas under discussion.

e. Participatory monitoring and evaluation is much enhanced by involving the community at the outset in the selection and recording of indicators of progress (Chapter 11).

Source: After Waters-Bayer, Bayer, and von Lossau 1995.

F
Livestock Credit Schemes

Livestock credit schemes were mentioned in Chapter 9 as offering scope for innovation. Aside from suggesting that this as an area that governments may wish to examine through a task force or consultancy, guidelines are constrained by lack of experience. The following outline of terms of reference shows the options that need to be explored.

Draft for a Study Team

Objective. A conceptual and operational plan is needed for a livestock credit scheme that will provide pastoralists with credit redeemable at a later date, in order to replace livestock relinquished during drought. It is envisioned that the credit will have a value expressed in livestock units rather than cash. Consideration will be given, however, to extending the scope of the scheme to allow credits to be acquired in ways other than relinquishing livestock at times of drought and to be used for purposes other than restocking. The scheme need not be run by government if higher standards of reliability and accountability can be assured using other operators.

Scope of study. Through consultation and data collection in government offices and pastoral districts, the study team should establish:

- the likely demand for livestock credit in relation to animal species, district, and drought intensity;
- potential uses for the animals acquired, considering grade of animal or carcass as well as species;
- any facilities needed for handling animals acquired or for ensuring a supply of replacement stock;
- the differential between the resale or carcass value of animals acquired and those to be bought as replacements;
- likely operating costs for the scheme given the expected throughput and price differential;
- the case for extending the scheme to fulfill additional functions;
- the desirability of operating more than one scheme, whether separated by region or function; and
- management options, including potential roles for government (central and local), herder organizations, the banking sector, NGOs, and other potential contributors.

Areas for detailed examination. Particular attention should be given to the following:

Issuance. One livestock credit would be issued for one live animal relinquished under drought conditions when the stock offered have a low value. What would be the implications of also issuing credits for:

- carcasses or hides/skins (perhaps two carcasses for one credit)?
- quality animals (perhaps plus cash payout)?
- traders to use instead of paying out in cash?
- others to use in poverty-focus restocking?

Redemption. Should the credit be redeemable:

- only for the same species as that relinquished?
- for other species (at what rate of exchange)?
- in the form of other goods, for example, services provided by herder organizations or others?
- just at government or community outlets?

Management. Should the whole scheme be operated by:

- one government department?
- government and a consortium of traders?
- local government, backstopped centrally?
- inviting bids from the private sector?

Logistics. The most problematic parts of the scheme are collecting and disposing of drought-stricken stock and ensuring stock for purchase at the time of redemption. What needs or opportunities exist for:

- mobile abattoirs for salvaging stock?
- herder organization inputs to restocking?
- other inputs to post-drought restocking?

These issues have numerous ramifications that require detailed analysis. In the case of mobile abattoirs, for example, the scope for private sector involvement will require in-depth analysis of (a) potential outlets for products (such as meat and bone meal, pet food, low-quality hides and skins), (b) commercial operators who might be interested and competent to participate, and (c) the form and extent of inducement or subsidy that may be necessary to attract competent operators.

G

Wildlife Management Projects

To ensure acceptance and long-term commitment by local people, wildlife management projects (WMPs) should ideally be initiated and executed by the rural communities themselves. WMPs should therefore aim to create an enabling environment (for example, adequate policies and local capacity building) within which local people appreciate the value of and derive significant benefits from wildlife resources.

Project Planning

Several elements should be considered when planning WMPs:

Distribution of responsibility and benefits is important. The objectives of the project and the limits of the authority that will devolve to the local people must be very clear. In general, it is best to leave the distribution of responsibilities and benefits to the community's own decisionmaking process.

The nature and value of wildlife resources affect both project direction and prospects for success. If the aim is to preserve whole ecosystems, potential for tourism development is critical. In the long-run, involving local participants may be the most effective and low-cost way to maintain a protected area. If the aim is sustainable consumptive use of wildlife, types and numbers of animals available, support capacity of the environment, and available markets are very important. Contracting with the private sector is likely to be the most effective way of mobilizing external support.

Local technical and institutional arrangements depend on the level of local participation envisioned.

- Participation in *benefits* will require a mechanism to allocate those benefits within the community and help ensure that rules are respected by community members.
- Participation in *decisionmaking* will require a different type of community organization, with training in technical aspects of wildlife ecology and management, communication, community organization, and motivation methods.
- Participation *in management and project execution* will require the highest level of training and preparation in technical, business, and policy aspects.

Communication is vital. In the planning phases, a clear understanding of local sociology and community dynamics is needed, while implementation requires a mechanism for interaction between all actors developed by trained people and institutionalized as a project component. In any WMP, environmental education should have a high priority.

Measures of Success

On the basis of present experience, success in WMPs should be measured by three criteria:

Do they advance our knowledge of how to implement community-based wildlife management, thus leading to better designed projects in the future?

WMPs are still experimental. So far, WMPs have (a) contributed to our understanding of how to pursue wildlife management with local participation; (b) identified the key elements of policy and the problems that can arise in implementation, and generated a variety of possible solutions; (c) demonstrated the value of wildlife resources and the possibility of rebuilding depleted populations; and (d) demonstrated the critical importance of understanding the sociological and economic factors that motivate rural people and govern their behavior, as well as of maintaining good communication channels. More needs to be known in all these areas and the potential for expanding WMPs into new situations.

Do or will WMPs stimulate the development of additional projects?

Several projects have stimulated follow-up activities, in some cases initiated by local communities seeking to participate in programs that they saw benefit their neighbors. More needs to be known about specific features that attract follow-up activity.

How well are WMPs achieving their stated objectives?

The direct objectives of WMPs commonly include one or more of the following:

Wildlife conservation. The lesson to date seems to be that partnership with local communities can be effective as long as benefits continue to flow to the people directly involved, and as long as they are clearly linked to

Box G-1. Lessons from CAMPFIRE

In 1993, the Zimbabwe CAMPFIRE program (Communal Areas Management Programme for Indigenous Resources) included twelve communal lands of more than 19,500 square kilometers. This communally owned land was officially designated to include wildlife in the resource-use and management system under the program. These districts manage wildlife with participation of the local population, obtain revenue from wildlife utilization, and distribute benefits to participating communities.

Two key reasons account for success. First, reduced macroeconomic distortions in Zimbabwe and a more competitive marketing system increased returns from wildlife. Second, local property rights assignment enabled financial rewards to be channeled back to local communities. Although revenue distribution remains controversial, CAMPFIRE has now issued guidelines that recommend a limit of 15 percent of revenue allocated to the district council and no more than 35 percent spent on management costs. Thus at least 50 percent of revenue can be passed to the community in the form of cash or project benefits. These guidelines were met in four of the eight districts for which data are available.

Four issues are still outstanding:

- District council authorities receive all the wildlife revenue and then allocate this revenue over a much wider population than that which is actually living with and conserving wildlife. A uniform distribution of the benefits between all areas is unlikely to result in benefits outweighing costs in the important producing areas.
- When wildlife revenue is used to fund social services and infrastructure projects in an area, the links between costs and benefits become weak.
- Appropriate authority is only granted to the district council, and local communities have yet to be given the opportunity to form effective resource management units. Devolving authority to village levels would reduce costs in wildlife management because villagers would have greater incentives to police themselves and report outside poachers.
- Training and assistance should be provided to develop project implementation skills at the community level.

CAMPFIRE has generated much debate over access to resources. With the migratory nature of wildlife, it is difficult to closely define producer communities, and even more difficult to ensure that those paying the highest costs receive the greatest benefits. Yet those who both live with and pay the costs of maintaining the wildlife must be made the primary beneficiaries. Both wildlife and tree resources are generally viewed as belonging to all Zimbabweans, but progress is being made in some districts toward recognizing that, for people in marginal areas, "their wildlife are their cattle."

Source: World Bank 1996.

conservation objectives. Whether it will be possible to pursue this approach on a larger scale remains to be tested.

Economic development based on sustainable wildlife utilization. WMPs that aim to mobilize wildlife resources as an economic asset for rural development should ultimately be measured in terms of returns on total investment, including the opportunity cost of labor and leisure. The revenue earning potential of wildlife means that financial returns can sometimes be quite high. Questions to be addressed include (a) whether the quest for higher returns will diminish the positive environmental impacts of conserving or managing wildlife in marginal lands, and (b) the relative merits of, for example, leasing safari hunting rights to private operators vs more complex and costly projects.

Local participation. The level of intended participation and how success in obtaining that objective is to be measured are usually not clearly defined. The connection between wildlife and benefits is not always clear because of confusion between community and individual inputs and rewards. Overall, there has been considerable progress in increasing the role and interest of local communities in conservation of wildlife resources, but the goal of ensuring local management of wildlife resources remains elusive.

Sustainability. Wildlife can offer an environmentally sustainable use of marginal lands. The main advantage is the possibility of increasing income without increasing pressure on the environment. Institutional sustainability of WMPs faces the same constraints as rural development generally: (a) the pursuit of short-term returns and rapid physical implementation and (b) the tendency to proceed with projects despite an unfavorable policy environment and lukewarm government commitment.

Source: Kiss 1990.

H
Monitoring Procedures

Chapter 11 is explicit about the types of information that should be collected in the course of process monitoring, but stopped short of specifying the precise methods to collect that information. This annex does not fully correct that omission, but it does explain why it arises.

There is no one set of ready-made methods that can be applied to process monitoring in all pastoral situations. When recording vegetation, for example, the configuration of transects and sampling intensity that would be appropriate for the subhumid zone would be wasteful and even deficient in recording ecological change in the arid zone. The same wastefulness and deficiency would arise from applying the recording methods appropriate to a trade-oriented pastoral system to a milk-subsistent one. Methods need to be selected to suit the specific ecology and pastoral system.

This specificity conflicts with the preference of institutions, the World Bank included, for indicators and methods of universal application. The three boxes that accompany this annex illustrate recent moves towards standardization in land quality indicators (LQIs), project performance indicators (PPIs), and rangeland monitoring (Boxes H-1, H-2, and H-3).

In these examples, standardization has not been taken to the point of specifying the actual parameters and methods to be used. Yet that specification is implied if the aim is standardization in the interests of comparing situations and responses to intervention.

The plea that is made here is that like should be compared with like. There is a danger in standardization of failing to recognize basic differences and drawing erroneous conclusions. All of the indicators listed in the boxes are relevant under some circumstances, but they mean different things as circumstances change.

In the case of LQIs (Box H-1), for example:

- The ratio of livestock to people means nothing unless related to a baseline situation, the sources of subsistence, and the species kept.
- The ratio of perennial to annual vegetation loses all significance in environments suited only to annual species.
- The rate of out-migration can (as in Yemen and northwestern Kenya) reflect cultural or environmental factors originating several generations ago.

Likewise, the indicators recommended for rangeland monitoring (Box H-3) are not all equally relevant under all conditions. Moreover their significance varies according to the time period over which any particular trend is observed. It seems not to be acknowledged that some sites can cycle through several states and forms of vegetation over time.

The predicament of which guidelines to follow can be resolved in several ways. One is to let consultants design and execute the monitoring component. But unless the consultants who define the work plan are excluded from executing it, chances are that unnecessarily labor-intensive methods will be proposed. The safer option is that proposed in Chapter 11 — forming a local task force to work with project management to ensure relevance in the monitoring program. The temptation to side-step formal monitoring in favor of *ad hoc* analysis during supervision and at mid-term should be strongly resisted. Subjective impressions are no substitute for objective time-series data collected during the life of a project.

The best way to ensure relevant monitoring — that is, to avoid collecting superfluous data with no management value while ensuring that critical data are not omitted — is to start with a zonal analysis. The manner in which climatic zone influences monitoring is illustrated in Table H-1. It is still necessary within that framework to examine all available sources on how cyclic and episodic events affect rangeland vegetation and NRM practices. And the monitoring of socioeconomic parameters must always be guided by the character of the pastoral systems and what changes are construed as constructive in the context of the development path being followed.

Box H-1. Land Quality Indicators (LQIs)

Land quality indicators (LQIs) developed by OECD within a framework of pressure-state-response (PSR) are currently being adapted for application in the World Bank.

The PSR framework links:

- pressures exerted on land by human activities,
- the resulting status of land resources, and
- the response of society to these changes.

Based on this framework, LQIs are grouped into:

- indicators of pressures upon lands,
- indicators of the state of the land quality and change in state over time, and
- indicators of responses by society to changes in land pressure and land quality.

LQIs vary according to the scale of application. As project indicators for use by task managers, LQIs relate closely to land management practices and processes in agroecosystems. They demonstrate changes over a short time period, and to ensure credibility need to be implemented in a standard manner in all projects. On a broad scale, they relate more directly to the policy arena. Subject to ongoing work in the World Bank, the following indicators are proposed for pastoral systems in arid zones:

Issues	*LQIs*
Pressure	
Population pressure on rangelands	Ratios of land/people and livestock/people
State	
Vegetation condition and cover	Ratio of perennial/annual vegetation, density of living perennial vegetation, ratio of vegetation biomass/feed demand
Vegetation quality	Ratio of palatable/unpalatable vegetation, ratio of young/mature perennial vegetation
Soil water storage capacity and runoff	Ratio of crusted soil surface area/total area
Response	
Response of land users	Rate of out-migration, range and quantity of products for sale, human diet, ratio of cereals/livestock products
Societal commitment	Budget for livestock and social services, number and cohesion of pastoral associations, number of conflicts over resources

Source: World Bank 1995c

Box H-2. Project Performance Indicators (PPIs)

Project performance indicators (PPIs) in use by the World Bank for the livestock sector cover input, output, and impact for three broad types of project activity. The following indicators have been proposed for range and pastoral development projects.

Objective	*Common inputs*	*Input indicator*	*Output indicator*	*Impact indicator*
Improve or at least maintain range production and condition	Range improvement investments using different techniques (deferred grazing, overseeding)	Costs for an area per improvement technique	Biomass production	Soil conditions, vegetation trends and animal offtake
Improve food security, especially to face droughts and enhance the quality of life	Establish early warning system and food and feed banks	Costs for early warning system and food/feed banks	Cereals/cattle prices, number of herders using the food/feed banks	Stability of household income, human nutrition status and health status
	Market and market information systems	Number of markets and type of systems to be established	Increased flexibility as shown by rate of change in herd size and degree of price stability	Share of unproductive animals in the herd and degree of herder participation in outside saving schemes
Create viable institutions	Organize pastoral associations	Number of associations to be created and pastoralists trained	Percentage of herders belonging to pastoral associations	Viability of associations, involvement in animal health, water and range management
	Assure adequate grazing rights	Institutional capability to register and enforce grazing rights	Area of exclusive grazing rights established	Number of internal and external conflicts involving herders, degree of assured access to key dry-season and water resources

Source: World Bank 1995b.

Box H-3. Revised Rangeland Monitoring

The United States National Research Council (NRC) has recommended a three-phase approach to assessing rangeland health, and has proposed criteria and methods to make that assessment. The rationale for a new approach is that while much research and experience supports the relationship of soil surface characteristics to rangeland health, basic knowledge of the effects of other soil properties is limited. Examples might include the effects of organic matter content and water holding capacity on nutrient cycling, energy flow, recovery mechanisms, and other aspects of rangeland health. The effect of grazing and other management practices on soil properties is also not well understood, therefore indicators of rangeland health are currently better developed for phase 1 than for phases 2 and 3. More effort is required to identify and test indicators of change in nutrient cycling, energy flow, and self-induced recovery mechanisms.

As is evident from Annex C and Box 10-1, there is a close parallel between the NRC approach and that already incorporated in the HRM model, which calls for diagnosis of the status of the four ecosystem blocks (succession, water cycles, mineral cycle, and energy flow) through comprehensive biological monitoring.

Phase	*Criteria*	*Indicators*
Soil stability and watershed function	Soil movement by wind and water	A-horizon depletion, rills and gullies, pedestaling, scour or sheet erosion, sedimentation or dunes
Distribution of nutrients and energy	Spatial and temporal distribution	Distribution of plants, litter distribution and incorporation, rooting depth and distribution, photosynthetic period
Recovery mechanism	Plant demographics	Age class distribution, plant vigor, germination and status of microsites

Source: NRC 1994.

Table H-1. Zonal Variation and Process Monitoring: the Effects of Relative Aridity on the Design of Monitoring Procedures

		Other arid zones		
Features to monitor	*Semi-desert zone*	*Annual grassland type*	*Perennial grassland type*	*Wetter zones*
Basic characteristics				
Location, environment, and vegetation	Fringing all major deserts: oppressively hot (with challengingly cold nights and winters as latitude and altitude increase); desiccating winds; vegetation concentrated in depressions and water courses, forming a mosaic of wooded or dwarf shrub grassland within predominantly barren land.	The Sahel of West Africa (between the Sahelo-Saharan and Sahelo-Soudanian belts) and more locally elsewhere (e.g., Dhofar highlands of Oman): less extreme climatically than semi-desert; seldom <150 mm rain per year but consistently long (9 month) dry seasons; relatively well-vegetated although with few perennial grasses.	Much of the subtropical and temperate drylands of the world (excluding deserts and their fringes) and the bimodal rainfall areas of tropical Africa: arid to semi-arid climate; not necessarily higher rainfall than the Sahel, but shorter or cooler dry seasons; vegetation mostly shrubland with (or with potential for) perennial grass.	Pastoralism is not much practiced in humid zones outside of Africa, but there it extends across much of the sub-humid zone of West Africa and locally into highlands: dry seasons can last 6-7 months but rainfall is relatively abundant; vegetation is tall-grass savanna or fire-induced grassland of various forms.
Rainfall in a 'good' wet season	Around 150 mm	250–300 mm	Bimodal: 200-250mm Monomodal: 300-400 mm	500–1,500 mm
Land use	Opportunistic seasonal grazing or year-round occupancy reliant on oases and/or camels.	Traditionally wet-season grazing, but increasingly sedentarized; grazing animals have difficulty surviving once the annual grasses start to disintegrate during the latter part of the dry season. Cropping very risky except on flood plains.	Pastoralism or (under pressure of population or because of favorable site conditions) agropastoralism; fewer nutritional constraints than the Sahel, although access to wetter zones is still a great help. Has important wildlife reserves.	Included on routes of transhumance in West Africa; dry season or drought reserve elsewhere; giving way increasingly to crop agriculture or locally to ranching.

Table H-1. (continued)

Features to monitor	Semi-desert zone	Other arid zones: Annual grassland type	Other arid zones: Perennial grassland type	Wetter zones
Environment	As a general rule, ecological monitoring is only useful if undertaken within a framework of ecological land units and accompanied by data on NRM. Key parameters for monitoring follow:			
Climate	Daily rainfall data (say, one site per 1,000 km^2), plus other relevant data, e.g., temperature and wind	Daily rainfall data (say, one site per 800 km^2), plus mistfall where relevant	Daily rainfall data (say, one site per 400 km^2), plus e.g., air humidity where fire is involved	Daily rainfall data (say, one site per 200 km^2), plus other data as relevant to crop production (evapo-transpiration, etc.)
Groundwater	Monitor abstraction/ recharge and water quality for each major groundwater province (sampling *every* well)	Usually sufficient to monitor just a sample of wells in each groundwater province	Same as for annual grassland type	Reduced sampling (±20% of wells) usually suffices
Springs	Monitor spring-fed oases	Monitor according to occurrence/importance	Same as for annual grassland type	Same as for drier zones, but with more springs to monitor
River flow	May be relevant, but requires monitoring of catchment outside the zone	Monitor seasonal discharge of all major rivers	Same as for annual grassland type	Monitor runoff and streamflow on a sample basis
Grass cover	Only relevant if perennials present or likely	Only relevant in flood plains or if obnoxious annuals likely	Monitor perennial cover and changes in grass flora	Same as for the arid perennial grassland type
Browse plants	Monitoring of overuse and regeneration is critical	Monitor overuse/regeneration of selected species	Same as for annual grassland type	Monitor also fire-tolerance

Table H-1. (continued)

Features to monitor	Semi-desert zone	Other arid zones: Annual grassland type	Other arid zones: Perennial grassland type	Wetter zones
Environment (continued)				
Bush density	Concentrate monitoring sites on flood plains and near settlements	Focus on settlement and where change in fire or grazing regimes is likely	Systematic monitoring required, recording all species and effects on grass growth	Same as for the arid perennial grassland type, including effects on tsetse
Soil erosion	Monitor dune formation	Monitor selected sites for accelerating erosion	Same as for annual grassland type	More intensive monitoring needed as rainfall and cropping increase
Fauna	Termite activity is as important as large fauna	Wildlife often sparse, selected species may warrant monitoring	Monitor biodiversity and selected species	Relate monitoring to the fauna remaining (if any)
Demography	Relate the baseline situation to the support capacity of the zone, and monitor change accordingly. Remember that diet (milk, meat, or grain-based) partly determines support capacity			
Population	Very low support capacity, intensive sampling needed	Relate to seasonality of use and nature of pastoral system.	Same as for annual grassland type	Much higher support capacity. Reliable data may derive from low sampling intensity
Diet	Monitor seasonality as well as longer term change	Same as for semi-desert	More intensive monitoring (greater biodiversity means that more bush foods are available)	Monitoring should allow for the fact that grain foods abundant and that quicker change in diets is likely
Health and education	Standards/expectations are relatively low, and services need to encourage mobility	Expectations somewhat higher; mobility still important	Standards and the need for mobile services will vary with pastoral system	Expectations much higher; quicker change likely; less need for mobile services
Out-migration	Monitor in the expectation that out-migration is a common drought strategy	Should be a common strategy; differentiate transhumant and settled communities	Often indicative of population pressure	Usually indicative of population pressure

Table H-1. (continued)

		Other arid zones		
Features to monitor	*Semi-desert zone*	*Annual grassland type*	*Perennial grassland type*	*Wetter zones*
Social structure	Kinship usually outweighs neighborliness	Similar to semi-desert (except where settlement is recent)	Neighborliness may outweigh kinship	Neighborliness likely to outweigh kinship
Leadership	A key parameter is strong leadership, essential to survival	Similar to semi-desert (except where settlement is recent)	Often many overlapping users: usually both tribal and neighborhood leadership need to be monitored	Customary institutions mostly superseded by civil administration
Retribution	Monitor that deterrents against incursion, etc. work, without being excessive	Same as for semi-desert	Same as for semi-desert	Expect civil procedures to be used more than customary sanctions
Conflict resolution	Access to water likely to be the overriding issue	Settlement and seasonality of access likely to be as important as water *per se*	Same as for annual grassland type, although with conflict priorities and procedures varying greatly between pastoral systems	Conflict with farmers likely to arise: customary solutions are still worth exploring in lieu of litigation
Politicization	Monitor whether politicization helps or hinders mobility and societal mechanisms	Focus on the extent to which transhumance is disadvantaged by the political voice of settlers	Same as for semi-desert, but expect major differences between systems	Usually already politicized: further monitoring may be unnecessary
Indebtedness	The cohesion of pastoral society depends to some extent on internalizing debt: the drier the zone, the more valid is the case for avoiding external debt	Give separate attention to attitudes in mobile and sedentary communities	Expect that attitudes to external debt will vary between pastoral groups	External debt usually accepted as a structural feature of society and the economy

Table H-1. (continued)

		Other arid zones		
Features to monitor	*Semi-desert zone*	*Annual grassland type*	*Perennial grassland type*	*Wetter zones*
Welfare support	Although part of 'social structure', societal welfare systems need separate monitoring. Because support mechanisms evolve with societies, they may show greater correlation with ethnicity than with zone			
Mechanisms	Large stock are likely to be treated as clan, not private property	Different mechanisms are likely among transhumants and (more recently) settled communities	The gifting of animals may be more prevalent here than in drier zones	More diversified mechanisms are likely because grain and employment are more readily available
Beneficiaries	'Need' likely to be defined broadly	Possibly restricted to drought and disease-loss victims	Expect major variation between systems (agropastoral vs pastoral) and with population pressure	Probably greater expectation that families in need find their own solutions
Exclusions	See above (but expect kinship to be influential)	Settled communities commonly seek to exclude transhumants	Incompetent pastoralists likely to be excluded from restocking	See above

Table H-1. (continued)

Features to monitor	*Semi-desert zone*	*Other arid zones*		*Wetter zones*
		Annual grassland type	*Perennial grassland type*	
NRM	The aspects considered below relate to societal, not household, NRM (see 'stockwealth,' below). Monitoring should clarify emergency strategies and give early warning of loss of mobility			
Territoriality	Defined in terms of river systems and escape routes	Was defined by transhumance itineraries; now often confused	Defined usually in terms of winter or dry season or cropped homelands and outlying grazing areas.	Usually involves inter-relationships and competition with agriculturalists
Mobility	Maximum flexibility required	Focus on dry-season strategy	Focus on (changing) rationales	Mobility less important
Reciprocity	Focus on arrangements for sharing key resource areas and accessing other zones	Same as for semi-desert	Relate sharing arrangements to the resources offered by each major ecological land unit	See above under 'Territoriality'
Drought reserve	Focus on flood plains and reserves outside the zone	Different approaches apply to wet-season and more permanent occupancy	Focus on making best use of available biodiversity	Give particular attention to the potential of croplands
Stockwealth and other indicators	Stockwealth is a key indicator. What constitutes a subsistence holding certainly varies among zones, but also depends on the uses to which livestock are put. It is an attribute that needs to be monitored at the household (not societal) level. Such attributes are excluded from this table. Institutional monitoring *is* undertaken at societal level, but is not much influenced by zone.			

Bibliography

Barker, J.S.F. 1992."Practical Issues for the Conservation and Improvement of Priority Breeds: General Considerations. *In* Hodges J. (ed), *The Management of Global Animal Genetic Resources*. Animal Production and Health Paper No. 104. Rome: FAO.

Bayer, W., and A. Waters-Bayer. 1994. "Forage Alternatives from Range and Field: Pastoral Forage Management and Improvement in the African Drylands." *In* Scoones, I. (ed), *Living with Uncertainty*. London: Intermediate Technology Publications.

Behnke, R.H., and C. Kerven. 1994. Redesigning for risk: tracking and buffering environmental variability in Africa's rangelands. *Natural Resource Perspectives* 1. London: Overseas Development Institute.

Behnke, R. H., I. Scoones, and C. Kerven (eds). 1993. *Range Ecology at Disequilibrium*. London: Overseas Development Institute/International Institute for Environment and Development/ Commonwealth Secretariat.

Bingham, S., and A. Savory. 1990. *Holistic Resource Management Workbook*. Covelo, California: Island Press .

Birgegärd, L. E. 1993 *Natural Resources Tenure: A Review of Issues and Experiences with Emphasis on Sub-Saharan Africa*. Rural Development Studies 31. Uppsala: Swedish University of Agricultural Studies, International Rural Development Centre.

Blackburn, H., H. Glimp, D. Phillipi, D. Child, and A. Pope. 1993. *Famine Mitigation Strategy Paper: Utilization and Preservation of Livestock Resources*. OFDA/ISDA Famine Mitigation Activity PASA. USAID Washington D.C.

Bojö, J. (ed). 1996. *The Economics of Wildlife: Case Studies from Ghana, Kenya, Namibia, and Zimbabwe*. AFTES Working Paper No. 19. Washington D.C.: World Bank.

Bromley, D. (ed). 1992. *Making the Commons Work*. San Francisco: Institute for Contemporary Studies.

Capezzuoli, S. 1994. *The Development of the Auto-Evaluation Process in the Gao and Timbuktu Regions in Northern Mali*. London: ACCORD.

Coppock, D. L. (ed). 1994. *The Borana Plateau of Southern Ethiopia: Synthesis of Pastoral Research, Development and Change, 1980-91*. ILCA Systems Study 5. Addis Ababa: International Livestock Centre for Africa (ILCA).

de Haan, C. 1994. *An Overview of the World Bank's Involvement in Pastoral Development*. Pastoral Development Network. Paper No. 36b. London: Overseas Development Institute.

Dregne, H., M. Kassas, and B. Rozanov. 1992. "Status of Desertification and Implementation of the United Nations Plan to Combat Desertification." *Desertification Control Bulletin* 20:6. New York: United Nations.

Gilles, J., and C. de Haan. 1994. *Recent Trends in World Bank Pastoral Development Projects: A Review of 13 Bank Projects in Light of the "New Pastoral Ecology"*. Pastoral Development Network. Paper No.

36b. London: Overseas Development Institute.

Gleick, J. 1993. *Chaos. Making a New Science*. London: Abacus. [An Abacus Book, first published in Cardinal by Sphere Books (and in London by Heinemann) 1988].

Glenn, J. 1988. *Livestock Production in North Africa and the Middle East: Problems and Perspectives*. World Bank Discussion Paper 39. Washington, D. C.:World Bank.

Hardin, G. 1968. The Tragedy of the Commons. *Science* 162:1243-1248.

IIED (International Institute for Environment and Development). 1994. "Special Issue on Livestock." RRA Notes No 20. London.

ILCA (International Livestock Centre for Africa). 1994. *1993 Annual Report*. Addis Ababa.

Kiss, A. (ed). 1990. *Living with Wildlife. Wildlife Resource Management with Local Participation in Africa*. World Bank Technical Paper 130. Washington D.C.: World Bank

Lane, C. 1996. *Pastures Lost. Barabaig Economy, Resource Tenure, and the Alienation of their Land in Tanzania*. ACTS Dryland Series 7. Nairobi: Iniatives Publishers (for the African Centre for Technology Studies [ACTS]).

Lane, C. and R. Moorehead. 1994. "New Directions in Rangeland and Resource Tenure and Policy." *In* Scoones, I. (ed). *Living with Uncertainty*. London: Intermediate Technology Publications.

Lühl, H-P. 1992. "Holistic Resource Management: Can it Beat Desertification?" *ILEIA Newsletter* 8:3. Leusden, Netherlands: Information Centre for Low-External-Input and Sustainable Agriculture.

Niamir, M. 1990. "Herders'Decision-Making in Natural Resource Management in Arid and Semi-Arid Africa." *Community Forestry Note 4*. Rome: FAO.

NRC (National Research Council). 1994. *Rangeland Health: New Methods to Classify, Inventory, and Monitor Rangelands*. Washington D.C.: National Academy Press.

Otzen, U. 1989. *Ecological Grazing Management. A Development Alternative for Brittle Environments — The Case of Namibia*. Berlin: German Development Institute.

Pratt, D.J. 1990. *Forms of Ranching and Pastoral Organization Appropriate for Isiolo District*. Government of Kenya, Emba-Meru-Isiolo Arid and Semi-Arid Lands Programme (EMI/ASAL), Isiolo Livestock Development Project. Nairobi: Government of Kenya.

Pratt, D.J. and M. Gwynne (eds). 1977. *Rangeland Management and Ecology in East Africa*. London: Hodder and Stoughton.

Sandford, S. 1983. *Management of Pastoral Development in the Third World*. Chichester: John Wiley & Sons.

______ 1981. *Review of World Bank Livestock Activities in Dry Tropical Africa*. Internal World Bank Document. Washington, D.C.: World Bank.

Savory, A. 1988. *Holistic Resource Management*. Washington D.C.: Island Press. [An African edition was published by Gilmour Publishing/HRM Publishers in 1991.]

Scoones, I. (ed). 1994. *Living with Uncertainty*. London: Intermediate Technology Publications.

Sere, C., H. Steinfeld, and J. Groenewold. 1996. "*World Livestock Production Systems — Current Status, Issues, and Trends*." Animal Production and Health Paper No. 127. Rome: FAO.

Shanmugaratnam, N., T. Vedeld, A. Mossige, and M. Bovin. 1992. *Resource Management and Pastoral Institution Building in the West African Sahel*. World Bank Discussion Paper 175. Washington, D.C.: World Bank.

Solomon Bekure, P. N. de Leeuw, B. E. Grandin, and P. J. H. Neate (eds). 1991. *Maasai Herding: An Analysis of the Livestock Production System of Maasai Pastoralists in Eastern Kajiado District, Kenya*. ILCA Systems Study 4. Addis Ababa: International Livestock Centre for Africa (ILCA).

Smuts, J. 1926. *Holism and Evolution*. London. [An American edition was reprinted in 1973 by Greenwood Press, Westport, Conn.]

Swift, J. 1994. "Dynamic Ecological Systems and the Administration of Pastoral Development." *In* Scoones, I. (ed) *Living with Uncertainty*. London: Intermediate Technology Publications.

Swift, J., and Abdi 1992. *Participatory Pastoral Development in Isiolo District: Socio-Economic Research in the Isiolo Livestock Development Project (ILDP)*. Isiolo, Kenya: ILDP Office.

Toulmin, C. 1994. *Gestion de terroirs: Concept and Development*. New York: UN Sudano-Sahelian Office.

Tyson, P. 1986. *Climate Change and Variability in Southern Africa*. Cape Town: Oxford University Press.

Umali, D., G. Feder, and C. de Haan. 1992. *The Balance Between Public and Private Sector Activities in the Delivery of Livestock Services*. World Bank Discussion Paper No. 163. Washington D.C.: World Bank.

Vergriette, B., and J. Rolland. 1994. "International Livestock Markets and Impact of Devaluation of CFA on West African Pastoral Economies." SOLAGRAL. Paper prepared for UNSO Donor/Specialized Agency Technical Consultation on Pastoral Development, 14-15 Feb. 1995, Eschborn, Germany.

Waters-Bayer, A., and W. Bayer. 1994. *Planning with Pastoralists: PRA and More. A Review of Methods Focused on Africa*. Eschborn, Germany: Deutsche Gesellschaft für Technische Zusammenarbeit (GTZ).

Waters-Bayer, A., W. Bayer, and A. von Lossau. 1995. *Participatory Planning with Pastoralists: Some Recent Experiences*. IIED Paper No. 58. London: IIED (International Institute for Environment and Development).

Wilson, R. T., P.N. de Leeuw, and C. de Haan (eds). 1983. *Recherches sur les systèmes des zones arides du Mali: résultats préliminaires*. CIPEA Rapport de recherche No 5. Addis Abeba: CIPEA (Centre International pour l'Elevage en Afrique

World Bank. 1995a. *North Africa and Iran. Rangelands Development in Arid and Semi-arid Areas : Strategies and Policies*. Washington D.C.

______ 1995b. "Performance Indicators in Bank-Financed Agricultural Projects." A First Edition Note (June). Washington D.C.

______ 1995c. "Workshop on Land Quality Indicators". June 20-21, 1995. Draft Concept Paper. Washington D.C.

______ 1995d. "World Bank Participation Sourcebook." Environment Department Papers. Washington D.C.

______ 1995e. "Decentralization: A New Strategy for Development". AGR Dissemination Notes No. 1, August. Washington D.C.

______ 1995f. "How Well Has Decentralization Worked for Rural Developement?" AGR Dissemination Notes No. 2, August. Washington D.C.

______ 1994. "Privatization of Animal Health Services. Improving Veterinary Efficiency." Agriculture Technology Notes No. 3, March.. Washington D.C.

User Guide

Deserts and rangelands account for the greater part of the earth's surface. Some rangelands are reserved for wildlife and others are allocated for ranching, but these areas are small compared with those occupied by communities of pastoralists.

Pastoralism is more than an agricultural sector activity. It is primarily a means of occupying territory. And like all forms of occupancy, pastoralism is diversified. A large part of the diversification is contained in the many roles which livestock play — as sources of subsistence and capital, in esthetics and social relationships, and in transport and trade. Consequently, the value of an animal in a pastoral economy is determined by the function that it fulfills, relative to the other animals that are available, which often bears little relationship to its market value.

Diversification also extends into the exploitation of mineral and other natural resources (including cropping, in the case of agropastoralists, and sometimes tourism), and often into artisanal activities or paid employment. Pastoralists who take paid employment or who provide goods or services to the market economy may do so as a step towards entering that economy — but often the primary motivation is to assure their identity in pastoral society. Thus, pastoral systems vary not just in the livestock species that are used and in their mobility and associated activities, but also in their basic economic orientation.

Development strategies need to be adapted to the orientation of the pastoral system(s) that are being addressed. The primary distinctions in orientation are between:

- maintaining a milk-based diet,
- trading in order to allow the regular consumption of grain foods, and
- trading for other purposes.

In addition, it is now commonplace for individuals of pastoral origin who are engaged in business or salaried employment to invest their savings in livestock kept in their home area. Those livestock then fulfill a very different role from the livestock that are contributing to family subsistence.

Population pressure and the distribution of stockwealth also greatly influence development strategy. The basic elements in pastoral development concern improvement in natural resource management and social services — but the goals, pathways, and strategies that are followed are all greatly affected by population pressure and the economic circumstances of individual households.

Moreover, the natural environment imposes constraints on what is feasible to accomplish in pastoral development. Mobile pastoralists cannot be settled and provided with 'luxuries' such as piped water, except in those few localities where the water resource is sufficient to support settlement. Even then, the livestock need to be kept mobile in order to avoid degrading land and starving the animals. Standards of living and food security appropriate to other rural situations are seldom attainable in the arid zone.

The implications are that while the vastness of the area and the acuteness of the need assure attention for arid zone pastoralism on the development agenda, there is need to be clear on the development paths that are being followed. Occasionally a project may focus exclusively on livestock marketing or wildlife utilization, or on relieving an emergency arising from drought or other calamity. But more usually projects will seek to bring to bear a cross-section of measures that aim to improve food security and the sustainability of NRM in a designated pastoral system.

Setting the scene has three requirements:

- information on demography and regional ecology, expressed spatially so that land use and physical infrastructure can be related to the availability of water supplies;
- a categorization of existing pastoral systems, differentiating communities and present practices by reference to mobility, livestock species and uses, associated enterprises, and relations with other economies; and
- the formulation of likely development paths, expressed in terms of the standards of living that are being sought, the uses to which water and other resources will be put, steps in the progression toward these goal, and the inputs required to get started.

The types of information required are set out on the following pages in the form of a check list. In most cases, the resulting project will be constructed around four elements, as elaborated in Part Two of this book:

1. Support for the establishment and operation of **herder organizations**, tailored to the tasks expected of the concerned communities.
2. Assurance of the **support and essential services** needed by community-based organizations, as provided by central and local government and the commercial sector.
3. Inputs specific to **drought management**, which is the essence of food security in arid lands. This may involve setting up early warning systems and strengthening response capabilities, centrally and locally, tailored to increasing intensities of drought. Certainly it will involve screening all **technical inputs** — water supplies, marketing facilities, and inputs to animal health and nutrition — to ensure that they contribute optimally to the mechanisms by which pastoralists currently cope with drought.
4. Establishment of a **monitoring system** that measures the impact of project interventions and non-project events in a manner than can guide successive steps in the development process. The monitoring must cover all aspects of NRM — the natural systems, the user system, and the geopolitical framework — and needs to focus on those indicators which, for the concerned site, provide the earliest and clearest indication of positive and negative change.

Guidelines specific to each of these areas are appended to the check list that follows.

Check List of Key Issues and Information for Planning Pastoral Development

Natural Environment

1. What is the mean annual rainfall?
a) <150 mm
b) 150-300 mm
c) 300-600 mm
d) 600-1,200 mm
e) >1,200

2. Over what period are rainfall data available?
a) <25 years
b) 25-50 years
c) 50-100 years
d) >100 years

3 Is there evidence of alternating wetter and drier cycles, each period lasting:
a) <7 years
b) 7-20 years
c) >20 years

4. Which months normally constitute the growing season?

5. Is the dormant season just dry or also subject to heavy frost or snow?

6. Does the ratio of rainfall to evapo-transpiration support perennial grass?
a) Consistently
b) During wetter cycles only

7. Is there evidence of cyclic shifts in vegetation type:
a) Between annual and perennial grasses
b) Through a more complex succession of woodland, grassland, and bushland

8. What is the distribution of wood vegetation?
a) Concentrated depressions and water courses
b) Widespread but mostly in one age class
c) Widespread in all age classes
d) Forming dense thickets
e) Depleted by overcutting for firewood

9. What is the extent of soil erosion?
a) Mild
b) Severe but localized
c) Requiring urgent attention

10. What are the main sources of drinking water?
a) Permanent lake or river
b) Deep groundwater
c) Shallow groundwater
d) Seasonal pans

11. What is the occurrence of wildlife?
a) Inconsequential
b) Important for tourism or other economic use
c) Minor hazard (e.g., by predators or biting flies)
d) Major impediment to livestock production

Pastoral System(s)

1. Is ethnicity simple or complex?

2. Are there several pastoral systems overlapping? On what is the variation based?
a) Ethnicity
b) Household circumstances
c) Differences in the livestock species held

3. Is there seasonal competition for resources?
a) For dry season or winter grazing
b) For wet season or summer grazing
c) For drought reserve
d) For any other specific resource

4. Is dependence on livestock (for most families) total or partial?
a) Livestock alone
b) Plus cropping (= agropastoralism)
c) Plus employment outside or pastoralism

5. Is change taking place rapidly or only gradually?
a) Situation relatively static
b) Rapid deterioration under population pressure
c) Changing expectations but little advance
d) Positive changes evident

6. How is stockwealth distributed?
a) Most families with at least a subsistence holding
b) Most families sub-subsistence

7. What is the dietary staple?
a) Milk
b) Meat
c) Home-grown grain
d) Purchased grain and additives

8. Is mobility constrained?
a) Freely mobile
b) Movement still possible but increasingly unwelcome
c) Recent loss of drought reserve or other key resource area
d) Enforced sedentarization
e) Feed supply adequate without movement

9. What drought management strategies (other than mobility) are employed?

10. How well do societal welfare mechanisms function?
a) Fallen into disuse along with other societal mechanisms
b) Rendered ineffectual through overuse
c) Effective for some (who, when, how?)

11. What is the level of dependence on famine relief and feed aid?
a) Rarely invoked
b) Regularly needed by poor families
c) Widely used in most years

Main uses of livestock (to be assessed separately for each pastoral system)

	Direct subsistence		*Draught/*	*Capital*		*Manure*	
Species	*Milk*	*Meat*	*transport*	*growth*	*Sale*	*Fuel*	*Fertilizer*
Camel							
Cattle							
Horse/mule							
Donkey							
Goat							
Sheep							

Enter *** to indicate principal use, ** for secondary use, and * for occasional use.

Institutional Framework

1. Does existing social territorial organization offer the institutions needed for community management in NRM and associated services?

2. Does government policy support pastoral development?
a) Encouraging mobility in arid zone pastoral systems
b) Recognizing customary law
c) Permitting decentralization

3. Are essential services funded and staffed appropriately?
a) With equitable funding (for the size of the pastoral population)
b) With staff who can communicate effectively with the pastoral population
c) With adequate mobility

4. Is local government effective?
a) Representative of the people
b) In receipt of adequate support from central government
c) Competent and poised to help herder organizations

5. Does land law recognize the needs and rights of pastoralists?
a) Confirming customary rights in homelands
b) Preventing land grabbing
c) Recognizing subsidiary rights of access

6. Does existing corporate law allow effective local organizations?
a) Managing facilities
b) Defending territory
c) Accessing credit
d) Dealing with state and commercial institutions

Distributors of World Bank Publications

Prices and credit terms vary from country to country. Consult your local distributor before placing an order.

ARGENTINA
Oficina del Libro Internacional
Av. Cordoba 1877
1120 Buenos Aires
Tel: (54 1) 815-8354
Fax: (54 1) 815-8156

AUSTRALIA, FIJI, PAPUA NEW GUINEA, SOLOMON ISLANDS, VANUATU, AND WESTERN SAMOA
D.A. Information Services
648 Whitehorse Road
Mitcham 3132
Victoria
Tel: (61) 3 9210 7777
Fax: (61) 3 9210 7788
E-mail: service@dadirect.com.au
URL: http://www .dadirect.com.au

AUSTRIA
Gerold and Co.
Weihburggasse 26
A-1011 Wien
Tel: (43 1) 512-47-31-0
Fax: (43 1) 512-47-31-29
URL: http://www .gerold.co/at.online

BANGLADESH
Micro Industries Development Assistance Society (MIDAS)
House 5, Road 16
Dhanmondi R/Area
Dhaka 1209
Tel: (880 2) 326427
Fax: (880 2) 811188

BELGIUM
Jean De Lannoy
Av. du Roi 202
1060 Brussels
Tel: (32 2) 538-5169
Fax: (32 2) 538-0841

BRAZIL
Publicacões Tecnicas Internacionais Ltda.
Rua Peixoto Gomide, 209
01409 Sao Paulo, SP.
Tel: (55 11) 259-6644
Fax: (55 11) 258-6990
E-mail: postmaster@pti.uol.br
URL: http://www .uol.br

CANADA
Renouf Publishing Co. Ltd.
5369 Canotek Road
Ottawa, Ontario K1J 9J3
Tel: (613) 745-2665
Fax: (613) 745-7660
E-mail: renouf@fox.nstn.ca
URL: http:// www .fox.nstn.ca/~renouf

CHINA
China Financial & Economic Publishing House
8, Da Fo Si Dong Jie
Beijing
Tel: (86 10) 6333-8257
Fax: (86 10) 6401-7365

COLOMBIA
Infoenlace Ltda.
Carrera 6 No. 51-21
Apartado Aereo 34270
Santafé de Bogotá, D.C.
Tel: (57 1) 285-2798
Fax: (57 1) 285-2798

COTE D'IVOIRE
Center d'Edition et de Diffusion Africaines (CEDA)
04 B.P. 541
Abidjan 04
Tel: (225) 24 6510;24 6511
Fax: (225) 25 0567

CYPRUS
Center for Applied Research
Cyprus College
6, Diogenes Street, Engomi
P.O. Box 2006
Nicosia
Tel: (357 2) 44-1730
Fax: (357 2) 46-2051

CZECH REPUBLIC
National Information Center
prodejna, Konviktska 5
CS – 113 57 Prague 1
Tel: (42 2) 2422-9433
Fax: (42 2) 2422-1484
URL: http://www .nis.cz/

DENMARK
SamfundsLitteratur
Rosenoerns Allé 11
DK-1970 Frederiksberg C
Tel: (45 31) 351942
Fax: (45 31) 357822

EGYPT, ARAB REPUBLIC OF
Al Ahram Distribution Agency
Al Galaa Street
Cairo
Tel: (20 2) 578-6083
Fax: (20 2) 578-6833

The Middle East Observer
41, Sherif Street
Cairo
Tel: (20 2) 393-9732
Fax: (20 2) 393-9732

FINLAND
Akateeminen Kirjakauppa
P.O. Box 128
FIN-00101 Helsinki
Tel: (358 0) 12141
Fax: (358 0) 121-4441
URL: http://www .akateeminen.com/

FRANCE
World Bank Publications
66, avenue d'Iéna
75116 Paris
Tel: (33 1) 40-69-30-56/57
Fax: (33 1) 40-69-30-68

GERMANY
UNO-Verlag
Poppelsdorfer Allee 55
53115 Bonn
Tel: (49 228) 212940
Fax: (49 228) 217492

GREECE
Papasotiriou S.A.
35, Stournara Str.
106 82 Athens
Tel: (30 1) 364-1826
Fax: (30 1) 364-8254

HAITI
Culture Diffusion
5, Rue Capois
C.P. 257
Port-au-Prince
Tel: (509 1) 3 9260

HONG KONG, MACAO
Asia 2000 Ltd.
Sales & Circulation Department
Seabird House, unit 1101-02
22-28 Wyndham Street, Central
Hong Kong
Tel: (852) 2530-1409
Fax: (852) 2526-1107
E-mail: sales@asia2000.com.hk
URL: http://www .asia2000.com.hk

INDIA
Allied Publishers Ltd.
751 Mount Road
Madras - 600 002
Tel: (91 44) 852-3938
Fax: (91 44) 852-0649

INDONESIA
Pt. Indira Limited
Jalan Borobudur 20
P.O. Box 181
Jakarta 10320
Tel: (62 21) 390-4290
Fax: (62 21) 421-4289

IRAN
Ketab Sara Co. Publishers
Khaled Eslamboli Ave.,
6th Street
Kusheh Delafrooz No. 8
P.O. Box 15745-733
Tehran
Tel: (98 21) 8717819; 8716104
Fax: (98 21) 8712479
E-mail: ketab-sara@neda.net.ir

Kowkab Publishers
P.O. Box 19575-511
Tehran
Tel: (98 21) 258-3723
Fax: (98 21) 258-3723

IRELAND
Government Supplies Agency
Oifig an tSoláthair
4-5 Harcourt Road
Dublin 2
Tel: (353 1) 661-3111
Fax: (353 1) 475-2670

ISRAEL
Yozmot Literature Ltd.
P.O. Box 56055
3 Yohanan Hasandlar Street
Tel Aviv 61560
Tel: (972 3) 5285-397
Fax: (972 3) 5285-397

R.O.Y. International
PO Box 13056
Tel Aviv 61130
Tel: (972 3) 5461423
Fax: (972 3) 5461442
E-mail: royil@netvision.net.il

Palestinian Authority/Middle East Index Information Services
P.O.B. 19502 Jerusalem
Tel: (972 2) 6271219
Fax: (972 2) 6271634

ITALY
Licosa Commissionaria Sansoni SPA
Via Duca Di Calabria, 1/1
Casella Postale 552
50125 Firenze
Tel: (55) 645-415
Fax: (55) 641-257
E-mail: licosa@ftbcc.it
Url: http://www .ftbcc.it/licosa

JAMAICA
Ian Randle Publishers Ltd.
206 Old Hope Road
Kingston 6
Tel: 809-927-2085
Fax: 809-977-0243
E-mail: irpl@colis.com

JAPAN
Eastern Book Service
3-13 Hongo 3-chome, Bunkyo-ku
Tokyo 113
Tel: (81 3) 3818-0861
Fax: (81 3) 3818-0864
E-mail: svt-ebs@ppp.bekkoame.or .jp
URL: http://www .bekkoame.or.jp/~svt-ebs

KENYA
Africa Book Service (E.A.) Ltd.
Quaran House, Mfangano Street
P.O. Box 45245
Nairobi
Tel: (254 2) 223 641
Fax: (254 2) 330 272

KOREA, REPUBLIC OF
Daejon Trading Co. Ltd.
P.O. Box 34, Youida
706 Seoun Bldg
44-6 Youido-Dong, Yeongchengpo-Ku
Seoul
Tel: (82 2) 785-1631/4
Fax: (82 2) 784-0315

MALAYSIA
University of Malaya Cooperative Bookshop, Limited
P.O. Box 1127
Jalan Pantai Baru
59700 Kuala Lumpur
Tel: (60 3) 756-5000
Fax: (60 3) 755-4424

MEXICO
INFOTEC
Av. San Fernando No. 37
Col. Toriello Guerra
14050 Mexico, D.F.
Tel: (52 5) 624-2800
Fax: (52 5) 624-2822
E-mail: infotec@rtn.net.mx
URL: http://rtn.net.mx

NEPAL
Everest Media International Services (P.)Ltd.
GPO Box 5443
Kathmandu
Tel: (977 1) 472 152
Fax: (977 1) 224 431

NETHERLANDS
De Lindeboom/InOr-Publikaties
P.O. Box 202
7480 AE Haaksbergen
Tel: (31 53) 574-0004
Fax: (31 53) 572-9296
E-mail: lindeboo@worldonline.nl
URL: http://www .worldonline.nl/~lindeboo

NEW ZEALAND
EBSCO NZ Ltd.
Private Mail Bag 99914
New Market
Auckland
Tel: (64 9) 524-8119
Fax: (64 9) 524-8067

NIGERIA
University Press Limited
Three Crowns Building Jericho
Private Mail Bag 5095
Ibadan
Tel: (234 22) 41-1356
Fax: (234 22) 41-2056

NORWAY
NIC Info A/S
Book Department
P.O. Box 6125 Etterstad
N-0602 Oslo 6
Tel: (47 22) 57-3300
Fax: (47 22) 68-1901

PAKISTAN
Mirza Book Agency
65, Shahrah-e-Quaid-e-Azam
Lahore 54000
Tel: (92 42) 735 3601
Fax: (92 42) 758 5283

Oxford University Press
5 Bangalore Town
Sharae Faisal
PO Box 13033
Karachi-75350
Tel: (92 21) 446307
Fax: (92 21) 4547640
E-mail: oup@oup.khi.erum.com.pk

Pak Book Corporation
Aziz Chambers 21
Queen's Road
Lahore
Tel: (92 42) 636 3222; 636 0885
Fax: (92 42) 636 2328
E-mail: pbc@brain.net.pk

PERU
Editorial Desarrollo SA
Apartado 3824
Lima 1
Tel: (51 14) 285380
Fax: (51 14) 286628

PHILIPPINES
International Booksource Center Inc.
1127-A Antipolo St.
Barangay, Venezuela
Makati City
Tel: (63 2) 896 6501; 6505; 6507
Fax: (63 2) 896 1741

POLAND
International Publishing Service
Ul. Piekna 31/37
00-677 Warzawa
Tel: (48 2) 628-6089
Fax: (48 2) 621-7255
E-mail: books%ips@ikp.atm.com.pl
URL: http://www .ipscg.waw.pl/ips/export/

PORTUGAL
Livraria Portugal
Apartado 2681
Rua Do Carmo 70-74
1200 Lisbon
Tel: (1) 347-4982
Fax: (1) 347-0264

ROMANIA
Compani De Librarii Bucuresti S.A.
Str. Lipscani no. 26, sector 3
Bucharest
Tel: (40 1) 613 9645
Fax: (40 1) 312 4000

RUSSIAN FEDERATION
Isdatelstvo <Ves Mir>
9a, Lolpachniy Pereulok
Moscow 101831
Tel: (7 095) 917 87 49
Fax: (7 095) 917 92 59

SINGAPORE, TAIWAN, MYANMAR, BRUNEI
Asahgate Publishing Asia Pacific Pte. Ltd.
41 Kallang Pudding Road #04-03
Golden Wheel Building
Singapore 349316
Tel: (65) 741-5166
Fax: (65) 742-9356
E-mail: ashgate@asianconnect.com

SLOVENIA
Gospodarski Vestnik Publishing Group
Dunajska cesta 5
1000 Ljubljana
Tel: (386 61) 133 83 47; 132 12 30
Fax: (386 61) 133 80 30
E-mail: belicd@gvestnik.si

SOUTH AFRICA, BOTSWANA
For single titles:
Oxford University Press Southern Africa
P.O. Box 1141
Cape Town 8000
Tel: (27 21) 45-7266
Fax: (27 21) 45-7265

For subscription orders:
International Subscription Service
P.O. Box 41095
Craighall
Johannesburg 2024
Tel: (27 11) 880-1448
Fax: (27 11) 880-6248
E-mail: iss@is.co.za

SPAIN
Mundi-Prensa Libros, S.A.
Castello 37
28001 Madrid
Tel: (34 1) 431-3399
Fax: (34 1) 575-3998
E-mail: libreria@mundiprensa.es
URL: http://www .mundiprensa.es/

Mundi-Prensa Barcelona
Consell de Cent, 391
08009 Barcelona
Tel: (34 3) 488-3492
Fax: (34 3) 487-7659
E-mail: barcelona@mundiprensa.es

SRI LANKA, THE MALDIVES
Lake House Bookshop
100, Sir Chittampalam Gardiner Mawatha
Colombo 2
Tel: (94 1) 32105
Fax: (94 1) 432104
E-mail: LHL@sri.lanka.net

SWEDEN
Wennergren-Williams AB
P. O. Box 1305
S-171 25 Solna
Tel: (46 8) 705-97-50
Fax: (46 8) 27-00-71
E-mail: mail@wwi.se

SWITZERLAND
Librairie Payot Service Institutionnel
Côtes-de-Montbenon 30
1002 Lausanne
Tel: (41 21) 341-3229
Fax: (41 21) 341-3235

ADECO Van Diermen EditionsTechniques
Ch. de Lacuez 41
CH1807 Blonay
Tel: (41 21) 943 2673
Fax: (41 21) 943 3605

TANZANIA
Oxford University Press
Maktaba Street
PO Box 5299
Dar es Salaam
Tel: (255 51) 29209
Fax: (255 51) 46822

THAILAND
Central Books Distribution
306 Silom Road
Bangkok 10500
Tel: (66 2) 235-5400
Fax: (66 2) 237-8321

TRINIDAD & TOBAGO, AND THE CARRIBBEAN
Systematics Studies Unit
9 Watts Street
Curepe
Trinidad, West Indies
Tel: (809) 662-5654
Fax: (809) 662-5654
E-mail: tobe@trinidad.net

UGANDA
Gustro Ltd.
PO Box 9997, Madhvani Building
Plot 16/4 Jinja Rd.
Kampala
Tel: (256 41) 254 763
Fax: (256 41) 251 468

UNITED KINGDOM
Microinfo Ltd.
P.O. Box 3
Alton, Hampshire GU34 2PG
England
Tel: (44 1420) 86848
Fax: (44 1420) 89889
E-mail: wbank@ukminfo.demon.co.uk
URL: http://www .microinfo.co.uk

VENEZUELA
Tecni-Ciencia Libros, S.A.
Centro Cuidad Comercial Tamanco
Nivel C2
Caracas
Tel: (58 2) 959 5547; 5035; 0016
Fax: (58 2) 959 5636

ZAMBIA
University Bookshop, University of Zambia
Great East Road Campus
P.O. Box 32379
Lusaka
Tel: (260 1) 252 576
Fax: (260 1) 253 952

ZIMBABWE
Longman Zimbabwe (Pte.)Ltd.
Tourle Road, Ardbennie
P.O. Box ST125
Southerton
Harare
Tel: (263 4) 6216617
Fax: (263 4) 621670

03/20/97